ESSENTIALS
OF
DRAFTING

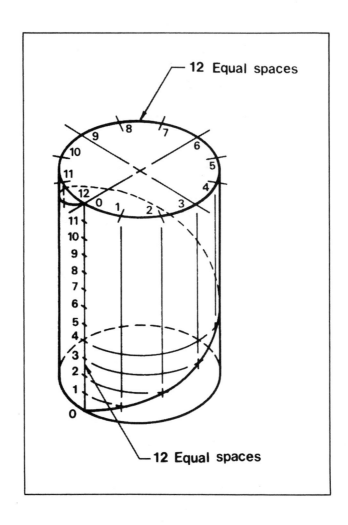

Second Edition

ESSENTIALS
OF
DRAFTING

JAMES D. BETHUNE

Boston University
College of Engineering
Boston, Massachusetts

PRENTICE-HALL, INC., Englewood Cliffs, New Jersey 07632

Library of Congress Cataloging-in-Publication Data

Bethune, James D.,
  Essentials of drafting.

  Bibliography:
  Includes index.
  1. Mechanical drawing. I. Title.
T353.B455 1987   604.2'4   85-30171
ISBN 0-13-284456-7

Editorial/production supervision and
  interior design: *Ellen Denning*
Cover design: *Whitman Studios, Inc.*
Manufacturing buyer: *Rhett Conklin*
Page Layout: *Meg Van Arsdale*

Printed in the United States of America

10 9 8 7 6 5 4 3 2 1

ISBN 0-13-284456-7 025

PRENTICE-HALL INTERNATIONAL (UK) LIMITED, *London*
PRENTICE-HALL OF AUSTRALIA PTY. LIMITED, *Sydney*
PRENTICE-HALL OF CANADA INC., *Toronto*
PRENTICE-HALL HISPANOAMERICANA, S.A., *Mexico*
PRENTICE-HALL OF INDIA PRIVATE LIMITED, *New Delhi*
PRENTICE-HALL OF JAPAN, INC., *Tokyo*
PRENTICE-HALL OF SOUTHEAST ASIA PTE. LTD., *Singapore*
EDITORA PRENTICE-HALL DO BRASIL, LTDA., *Rio de Janeiro*

To KENDRA

CONTENTS

4 PROJECTION THEORY   81

5 THREE VIEWS OF AN OBJECT   97

6 DIMENSIONS AND TOLERANCES   126

7   OBLIQUE SURFACES AND EDGES   166

8   CYLINDERS   199

9   CASTINGS   224

PREFACE
TO THE SECOND EDITION

The second edition of this book is a continuation of the ideas and teaching principles presented in the initial edition. Some areas have been expanded, others rewritten, but the book is still aimed at the student who is taking just one course in drafting and does not intend to become a drafter.

The step-by-step problem-solving format accompanied by many illustrations has been maintained. Each subject is covered in theory, then supported by sample applications. Students should be able to do the exercise problems given at the end of the chapter by refering to the text and to the examples within the chapter.

Approximately 90 new exercise problems have been added. They serve to emphasize my belief that drafting is best learned by doing lots of drawings. Many of the new exercise problems are in metrics.

Extensive changes have been made to Chapters 6, 14, and 15. Chapter 6, Dimensions and Tolerances, has more examples of dimensioning techniques and a greatly expanded section on how to call out and interpret tolerances.

Chapter 14, Production Drawings, has been completely rewritten. The relationship between assembly drawings, detail drawings, and parts lists is explained and demonstrated. Also included is an explanation of how drawing revisions and drawing notes are placed and referenced on a drawing. Several new exercise problems have been added to allow students to apply the material.

Chapter 15, Isometric Drawings, tells how to draw fasteners in isometric. The discussion covers hexagon-shaped heads and nuts, as well as washers. There are five new exploded-drawing problems that should interest both the average and advanced student.

I would like to thank the many who took the time to send comments, corrections, and suggestions. They were most helpful. I encourage you to continue to write to me via Prentice-Hall, with other comments. I really do appreciate them.

I would also like to thank Tim McEwen, my editor at Prentice-Hall, and Ellen Denning, the production editor, for their help and guidance. I would again like to thank Professor George Cushman of Wentworth Institute of Technology, whose suggestions for the first edition continue to be of value.

JAMES D. BETHUNE

College of Engineering
Boston University

1

DRAFTING TOOLS AND THEIR USE

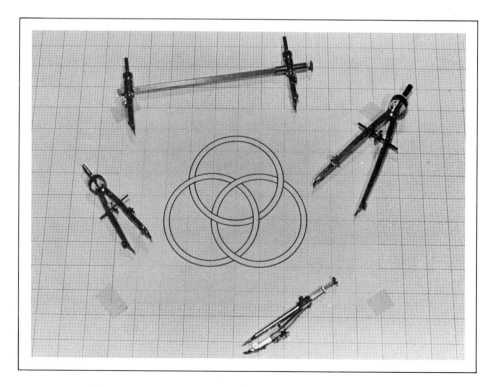

FIGURE 1-0

1-1 INTRODUCTION

This chapter explains and demonstrates how to use basic drafting tools. Most sections in the chapter are followed by exercises especially designed to help you develop skill with the particular tool being presented. Try each tool immediately after reading about it by doing the appropriate exercises. As you work, try to learn the capabilities and usage requirements of each tool, because it is important that you know how to use each tool with technical accuracy, skill, and creativity.

1-2 PENCILS, LEADHOLDERS, AND ERASERS

Figure 1-1 shows several different pencils and leadholders. Most drafters prefer to draw with leadholders instead of pencils because leadholders maintain a constant weight and balance during use which makes it easier to draw uniform lines.

Regardless of whether a leadholder or a pencil is used, its lead

FIGURE 1-1   Pencils and leadholders.

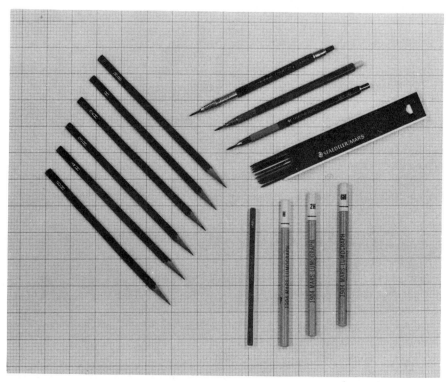

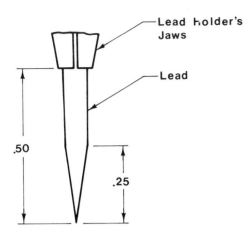

Lead holder's Jaws

Lead

.50

.25

FIGURE 1-2   Shape of a properly sharpened lead.

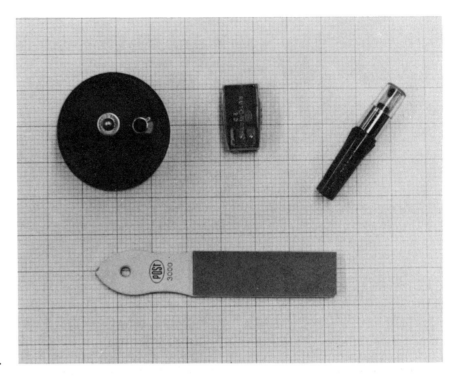

FIGURE 1-3   Lead sharpeners.

must be kept sharp with a tapered, conical point like the one shown in Figure 1-2. Figure 1-3 shows several different lead sharpeners and Figure 1-4 shows how to sharpen a lead by using a sandpaper block.

When sharpening a lead, care should be taken to keep the graphite droppings away from the drawing. Most drafters keep a cloth or piece of clay handy to wipe the excess graphite from a newly sharpened lead.

Leads come in various degrees of hardness, graded H to 9H. The higher the number, the harder the lead. Light layout and projection lines are usually drawn with the harder leads; darker lines, used for detailing and lettering, are drawn with the softer leads.

Figure 1-5 shows several different kinds of erasers and an erasing shield. The harder erasers are used for removing ink lines and the softer ones are used for removing pencil lines. Gum erasers (very soft) are used when large amounts of light erasing are required.

An erasing shield enables a drafter to erase specific areas of a drawing and thereby prevents excessive redrawing of lines that might otherwise have been erased. To use an erasing shield, place it on the drawing so that the area to be removed is exposed through one of the cutouts. (The various cutouts are shaped to match common drawing configurations.) Hold the shield down firmly and rub an eraser into the aligned cutout until the desired area is removed. When the erasing is finished, the excess eraser particles should be brushed off. Figure 1-6 demonstrates the method.

FIGURE 1-4   Sharpening a lead using a sandpaper block.

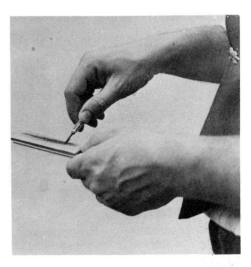

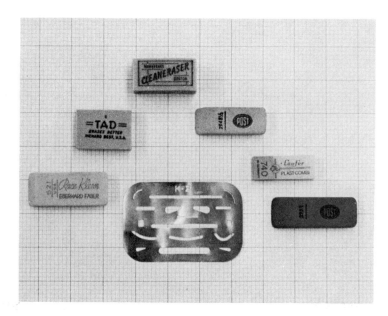

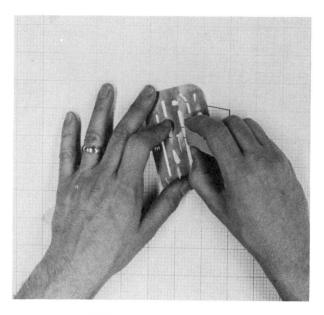

FIGURE 1-5   Erasers and an erasing shield.

FIGURE 1-6   Using an erasing shield.

1-3   SCALES

Scales are used for linear measuring. Figure 1-7 shows a grouping of several different kinds of scales. The scale most commonly used by drafters is one with its inches graduated into 16 divisions with each division measuring one-sixteenth of an inch. Figure 1-8 shows part of a "16-to-the-inch" scale along with some sample measurements. Unlike a real scale, the scale in Figure 1-8 has the first inch completely labeled to help you become familiar with the different fractional values. Measurements more accurate than one-sixteenth must be estimated. For example, 1/32 is halfway between the 0 and the 1/16 marks.

Figure 1-9 shows part of a decimal scale. Each inch is divided into 50 equal parts making it possible to make measurements within 0.01

FIGURE 1-7   Scales.

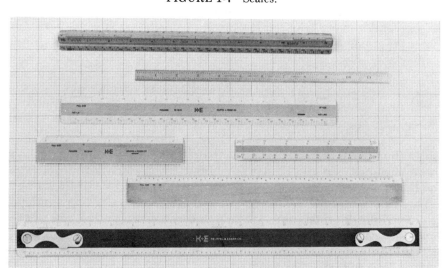

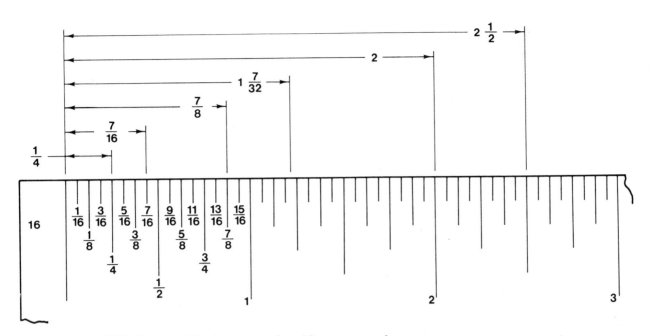

FIGURE 1-8 A 16-to-the-inch scale with some sample measurements.

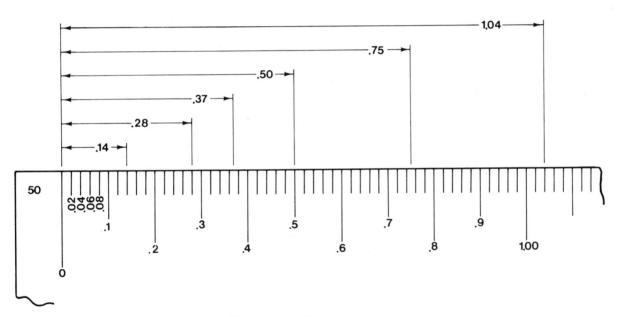

FIGURE 1-9 Decimal scale with some sample measurements.

inch (hundredth of an inch) accuracy. Several sample readings have been included and the first 0.10, unlike a real decimal scale, has each graduation mark labeled.

Many scales are set up for other than full-sized drawing. For example, the 1/2 scale enables a half-sized drawing to be made directly without having to divide each dimensional value by 2. Three-quarter scales enable direct 3/4-sized drawings to be made, and so on.

All fractional scales are read as shown in Figure 1-10. Only one of

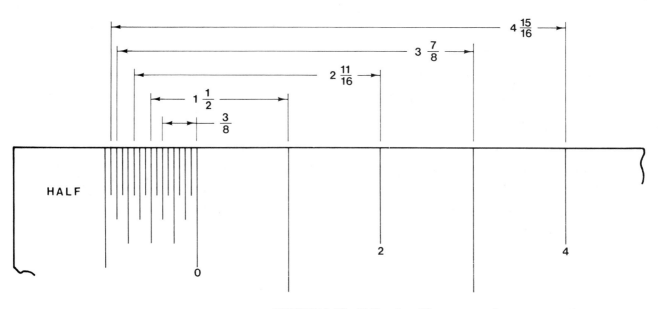

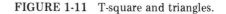

FIGURE 1-10   Half scale with some sample measurements.

the sections representing an inch is graduated into fractional parts. This graduated section is located to the left of the "0" mark. When making a reading (for example, 3-7/8) on a fractional scale, read the whole (3) part of the number to the right of the "0" and the fractional part (7/8) to the left. See Figure 1-10 for an example of a 3-7/8 reading on a half scale. See Sections 13-2, 13-3, and 13-4 for metric measurements.

1-4   T-SQUARE AND TRIANGLES

A T-square is used as a guide for drawing horizontal lines and as a support for triangles which, in turn, are used as guides for drawing vertical and inclined lines. Figure 1-11 shows a T-square and several different sizes and types of triangles, including an antique wooden one.

FIGURE 1-11   T-square and triangles.

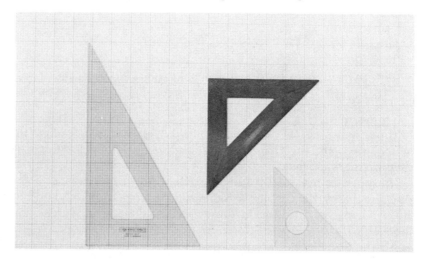

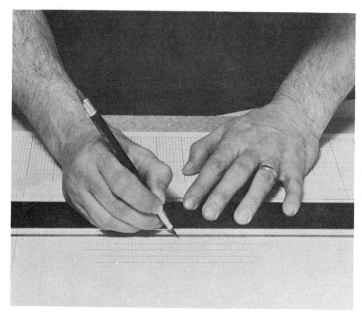

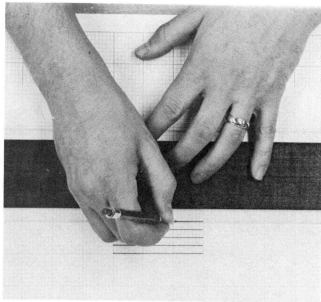

a.

b.

FIGURE 1-12   Drawing a horizontal line using a T-square
as a guide.

To use a T-square or triangle as a guide for drawing lines, hold the
pencil as shown in Figure 1-12 and pull the pencil along the edge of the
straight edge from left to right. (These instructions are for right-handed
people. Left-handed people should reverse these directions.) Rotate
the pencil as you draw so that a flat spot will not form on the lead. Flat
spots cause wide, fuzzy lines of uneven width. Always remember to
keep your drawing lead sharp.

When using a T-square, hold the head (top of the T) firmly and
flat against the edge of the drawing board. Use your left hand to hold
the T-square still and in place while you draw. When you move the T-
square, always check to see that the head is snug against the edge of the
drawing board before you start to draw again.

When a T-square and a triangle are used together to create a guide
for drawing, the left hand must not only hold the T-square in place; it
must also hold the edge of the triangle firmly and flat against the edge
of the T-square. To accomplish this, use the heel of your hand to hold
the T-square in place and your fingers to keep the triangle against the
T-square (see Figure 1-13).

It is important that all your tools be accurate. A T-square, for
example, must have a perfectly straight edge. If it does not, you will
draw wavy lines and inaccurate angles with the triangles. To check a
T-square for accuracy, draw a long line by using the T-square as a guide.
Then flip the T-square over, as shown in Figure 1-14, and, using the
same edge you just used as a guide, see if the T-square edge (now upside-
down) matches the line. If it does not, the T-square is not accurate.

Triangles should be checked for straightness in the same manner
used to check a T-square, but, in addition, they must be checked for
"squareness." To check a triangle for squareness, align the triangle
against the T-square and draw a line by using the edge of the triangle
which forms a 90° angle to the T-square as a guide. Holding the T-
square in place, flip the triangle over, as shown in Figure 1-15, and see

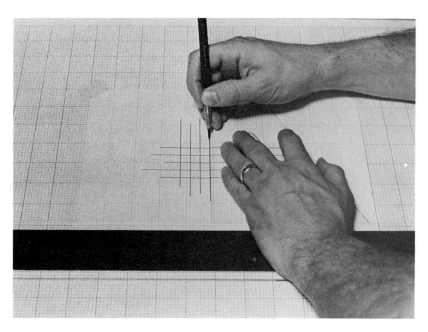

FIGURE 1-13 Drawing a vertical line using a T-square and a triangle as a guide.

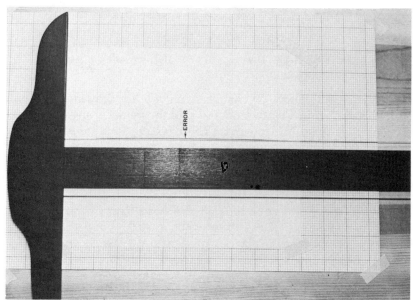

FIGURE 1-14 Measuring the following angles.

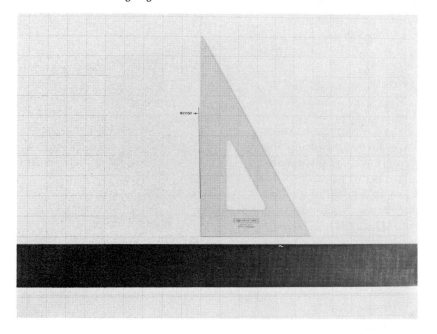

FIGURE 1-15 Checking a triangle for squareness.

FIGURE 1-16   Using a T-square and a triangle as a guide
for drawing parallel lines.

if the triangle edge matches the line. If it does not, the triangle is not
square, meaning either that the 90° angle is not 90°, or that the edge of
the triangle is curved, or that the edge of the T-square is curved.

 To use the T-square and triangle as a guide for drawing a line parallel to a given inclined line, align the long leg of the triangle with the
given line and then align the T-square to one of the other legs of the
triangle, as shown in Figure 1-16. By holding the T-square in place with
your left hand, you can slide the triangle along the T-square and the
long leg will always be parallel to the originally given line. You may
substitute another triangle in place of the T-square, as shown in Figure
1-17, and obtain the same results. Note that in either setup, the short
leg of the moving triangle is 90° to the long leg, meaning that you have

FIGURE 1-17   Using two triangles as a guide for drawing
inclined parallel lines.

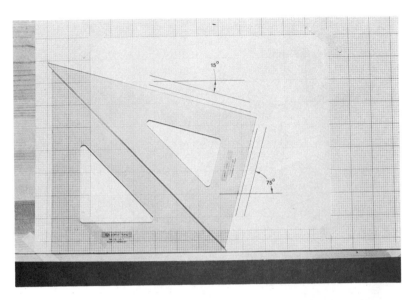

FIGURE 1-18 Using a T-square, a 45–45–90 triangle, and a 30–60–90 triangle to draw lines 15° and 75° to horizontal.

a guide not only for parallel lines, but also for lines perpendicular to those parallel lines.

A T-square may be used in combination with a 30–60–90 triangle and a 45–45–90 triangle to produce a guide for drawing lines which are 15° and 75° to the horizontal. Figure 1-18 illustrates how this is done.

1-5   COMPASS

A compass is used to draw circles and arcs. The three basic kinds of compasses are drop, bow, and beam. The bow is the most common (see Figure 1-19).

To use a compass, set the compass opening equal to the radius of the desired circle or arc by using a scale as shown in Figure 1-20. Then

FIGURE 1-19 Compasses.   **FIGURE 1-20** Setting a compass.

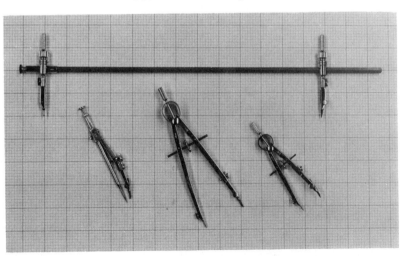

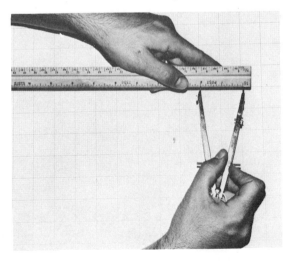

place the compass point directly on the circle center point and, using only one hand as shown in Figure 1-21, draw in the circle.

A compass lead must be sharpened differently from a pencil lead since the compass lead cannot be rotated during use to prevent flat spots from forming. Figure 1-22 shows how to sharpen a compass lead and Figure 1-23 shows a close-up of a properly sharpened compass lead.

a.

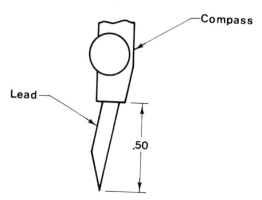

FIGURE 1-21 Drawing with a compass. (a) Courtesy of Teledyne Post, Des Plaines, IL 60016.

b.

FIGURE 1-22 Sharpening a compass lead.

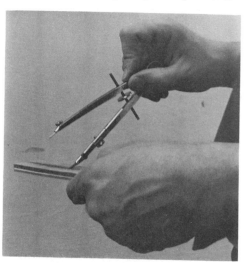

FIGURE 1-23 Shape of a properly sharpened compass lead.

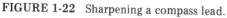

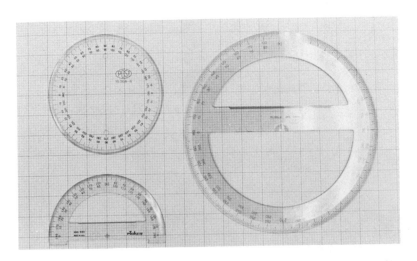

FIGURE 1-24   Protractors.

1-6   PROTRACTORS

A protractor is used to measure angles. Figure 1-24 shows three different kinds of protractors. The edge of a protractor is calibrated into degrees and half degrees. Figure 1-25 shows part of a typical protractor edge along with some sample measurements. Measurements more accurate than half a degree $(0.5°)$ must be estimated.

To measure an angle, place the center point of the protractor on the origin of the angle so that one leg of the angle aligns with the $0°$ mark on the protractor. Read the angle value where the other leg of the angle intersects the calibrated edge of the protractor.

FIGURE 1-25   Protractor with some sample measurements.

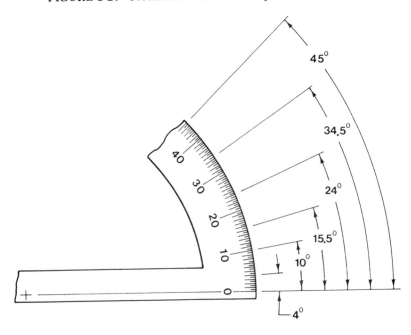

FIGURE 1-26   Curves.

1-7   CURVES

Curves are used to help draw noncircular curved shapes. Drafters refer to them as *French curves* or *ship's curves*, depending on their shapes (ship's curves look like the keel of a ship). Figure 1-26 shows a grouping of curves.

Noncircular shapes are usually defined by a series of points and a curve is used to help join the points with a smooth, continuous line. Using a curve to help create a smooth line is difficult and requires much practice. Most students make the error of trying to connect too many points with one positioning of the curve. Figure 1-27 shows a series of points that are partially connected. The curve is in position to serve as a guide for joining *only* points 3 and 4 — not 3, 4, and 5 — even though all three seem to be aligned. To join point 5 using the shown curve position would make it almost impossible to draw a continuous smooth curve.

FIGURE 1-27   Aligning a curve with given points.

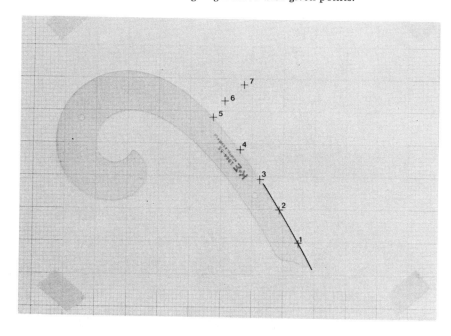

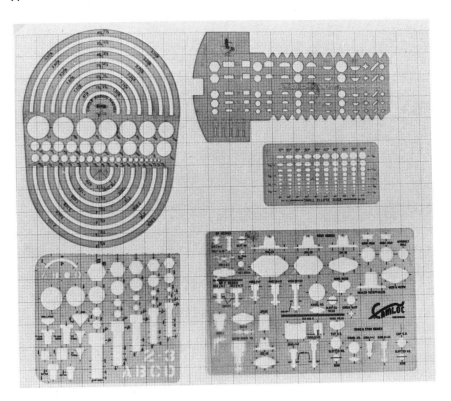

FIGURE 1-28   Templates.

1-8   TEMPLATES

Templates are patterns cut into shapes useful to a drafter. They save drawing time by enabling the drafter to accurately trace a desired shape. Some templates provide shapes that are difficult to draw with conventional drawing tools (very small circles, for example). Other templates provide shapes that would be tedious and time consuming to lay out and draw (ellipses, for example). Figure 1-28 shows a sampling of templates.

The most common template used in mechanical drafting is the circle template (see Figure 1-29). The holes of a circle template are

FIGURE 1-29   Circle templates.

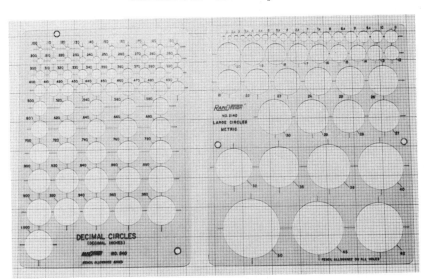

FIGURE 1-30   Using a circle template.

labeled by diameter size and are generally made slightly oversized to allow for lead thickness. Always check a circle template before initial use to see if lead allowance has been included.

To use a circle template, locate the center point of the future circle with two lines 90° to each other. Align the template with the two 90° lines by using the four index marks printed on the edge of the template hole. Draw in the circle. Keep the leadholder vertical and constantly against the inside edge of the hole pattern. Check the finished circle with a scale. Figure 1-30 shows how to use a circle template.

1-9   OTHER TOOLS

There are many tools, other than the ones already presented, which are used to help create technical drawings. Figure 1-31, for example, shows an adjustable curve ("snake") which is very helpful when drawing unusually shaped curves. Figure 1-31 also shows several other tools.

Figure 1-32 illustrates a drafting machine. A drafting machine is a combination T-square, triangle, protractor, and scale which, when used properly, will greatly increase drawing efficiency. The information previously presented for using a T-square, triangle, protractor, and scale may be directly applied to using a drafting machine. Check the manufacturer's instructions for the specific functions of the machines.

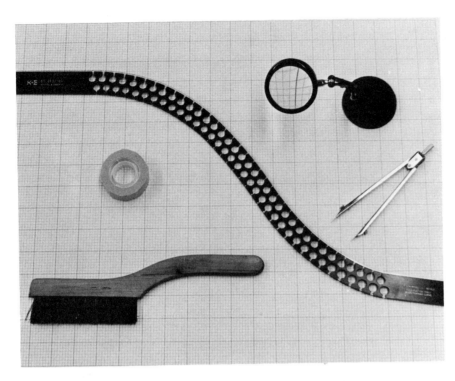

FIGURE 1-31 Magnifying glass, dividers, adjustable curve (snake), tape, and brush.

FIGURE 1-32 Using a drafting machine (Courtesy of Teledyne Post, Des Plaines, IL 60016.)

PROBLEMS

P1-1   Measure the lines shown in Figure P1-1(a) to the nearest 1/16 inch; (b) to the nearest 0.01 inch.

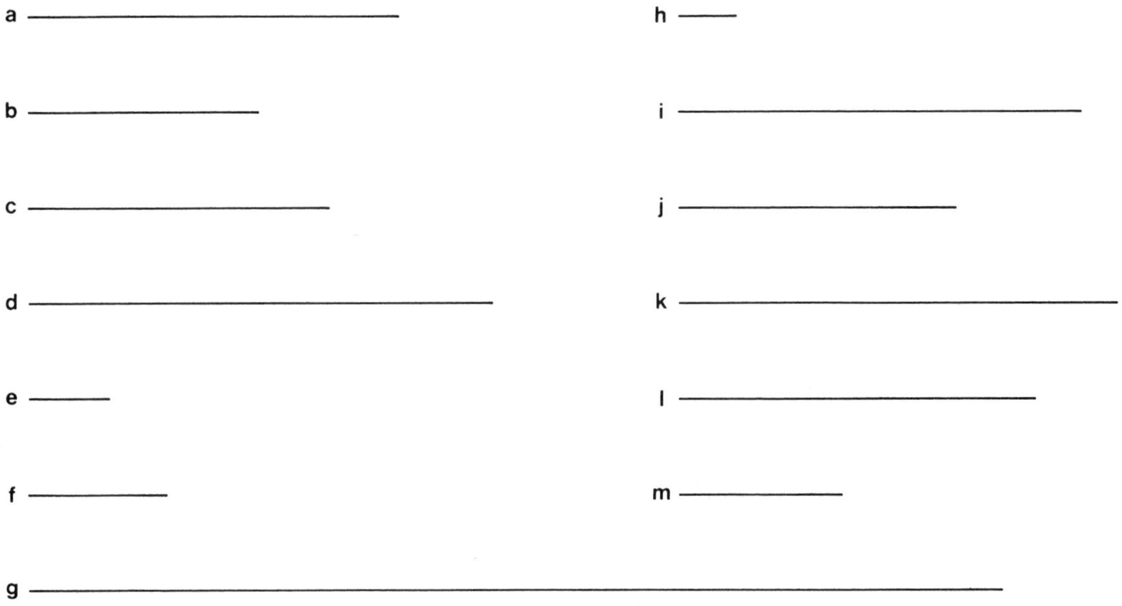

a ————————————————

h ——

b ————————————

i ————————————————————

c ——————————————

j ————————————————

d ——————————————————————

k ——————————————————————

e ————————

l ——————————————————

f ——————————

m ——————————————

g ————————————————————————————————————

FIGURE P1-1

P1-2   Draw four 8-inch lines as shown in Figure P1-2. Make the
lines very light and very thin. Define the left end of each
line as point 1.

Using point 1 as a starting point, measure off and label the
following points:

Line 1	Distance	Value
	1-2	1-3/8
	1-3	1-9/16
	1-4	2-1/4
	1-5	2-7/8
	1-6	3-21/32
	1-7	5
	1-8	5-3/8
	1-9	5-3/4
	1-10	6-5/16
	1-11	7-15/16

Line 2	Distance	Value
	1-2	0.8
	1-3	1.4
	1-4	2.6
	1-5	3.1
	1-6	4.3
	1-7	5.0
	1-8	5.5
	1-9	6.2
	1-10	6.9
	1-11	7.7

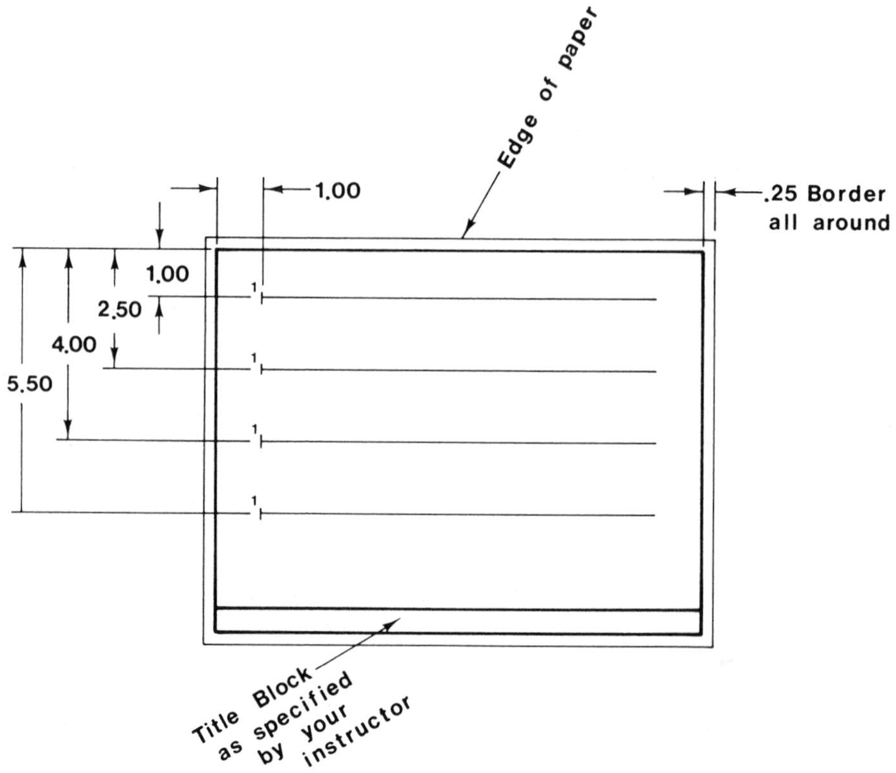

FIGURE P1-2

Line 3

Distance	Value
1-2	0.38
1-3	1.25
1-4	2.44
1-5	3.06
1-6	4.22
1-7	5.00
1-8	5.50
1-9	6.13
1-10	6.94
1-11	7.88

Line 4

The values given must be reduced by a factor of 2 to fit on the line. Therefore, using a 1/2 scale, draw the values.

Distance	Value
1-2	1.75
1-3	2.25
1-4	4.38
1-5	6.75
1-6	8.63
1-7	10.00
1-8	11.50
1-9	12.88
1-10	13.75
1-11	16.00

P1-3   Measure the lettered distances shown in Figure P1-3 to the
nearest 1/16 inch.

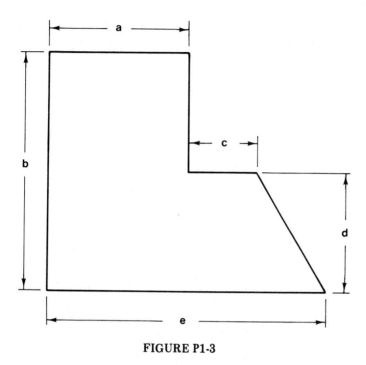

FIGURE P1-3

P1-4   Measure the lettered distances shown in Figure P1-4 to the
nearest 0.01 inch.

FIGURE P1-4

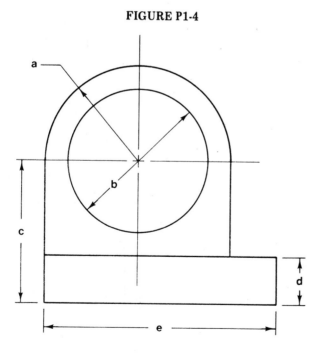

P1-5   Redraw each figure.
through
P1-10

FIGURE P1-5

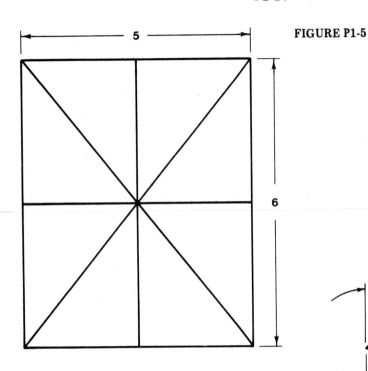

FIGURE P1-6

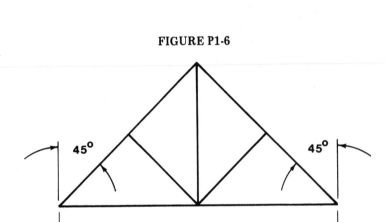

FIGURE P1-7

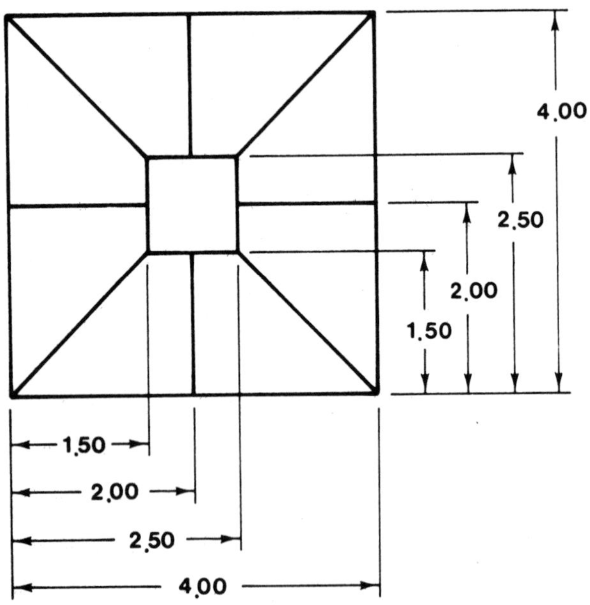

FIGURE P1-8

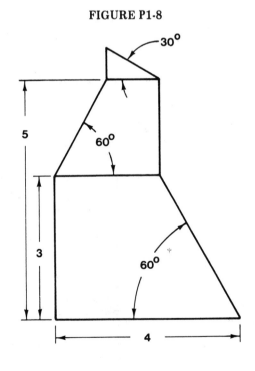

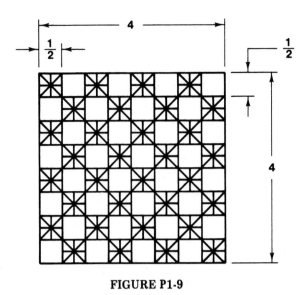

FIGURE P1-9

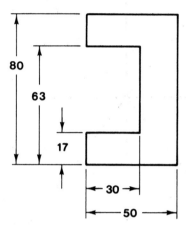

FIGURE P1-10 All dimensions are in millimeters.

P1-11   Redraw Figure P1-11. Use only a T-square, a 30–60–90 triangle, and a 45–45–90 triangle or a drafting machine.

FIGURE P1-11

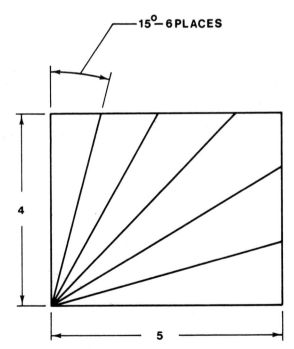

P1-12   Redraw Figure P1-12. The smallest circle is 1 inch in diameter and each additional circle is 1 inch larger in diameter up to 6 inches.

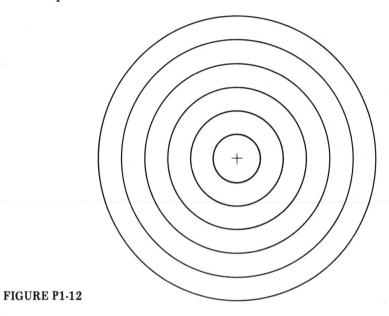

FIGURE P1-12

P1-13   Redraw Figure P1-13.

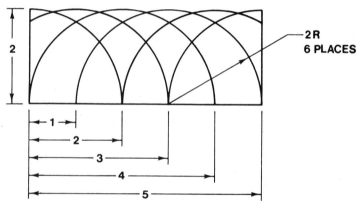

FIGURE P1-13

P1-14   (a) Redraw Figure P1-14(a); (b) measure the angles in Figure P1-14(b).

FIGURE P1-14(a)

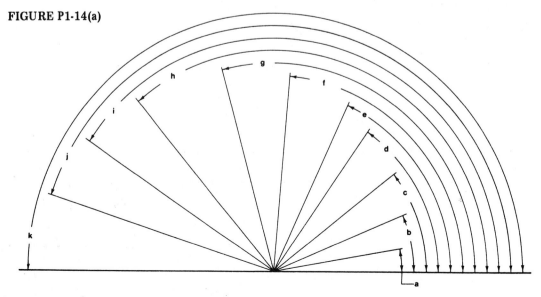

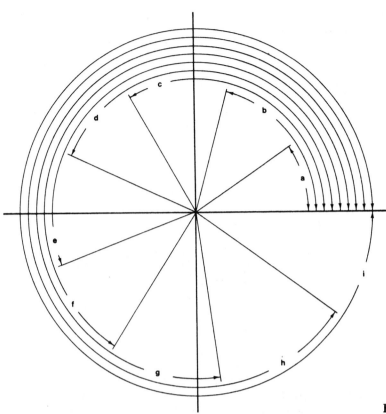

FIGURE P1-14(b)

P1-15 Redraw each figure.
through
P1-17

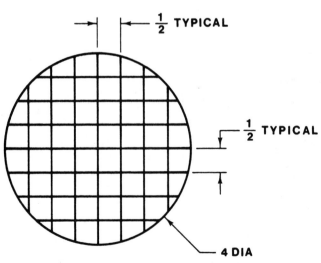

FIGURE P1-15

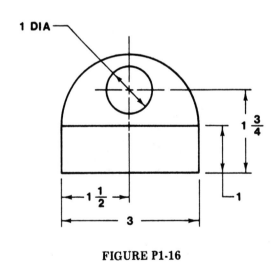

FIGURE P1-16

FIGURE P1-17

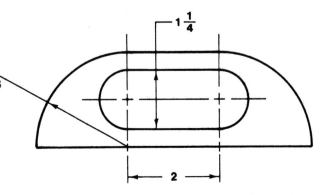

P1-18 Redraw Figure 1-18. Use a circle template to draw in the rounded corners.

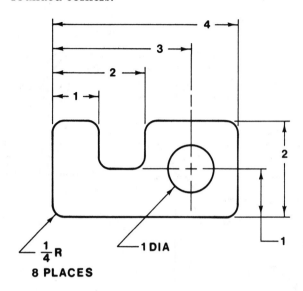

FIGURE P1-18

P1-19 Draw two curves x versus y_1 and x versus y_2 using the data points provided in Figure P1-19(a). Use an axis system like the one presented in Figure P1-19(b) and carefully label each curve.

FIGURE P1-19

a.

x	y_1	y_2
.00	.00	2.00
.50	1.00	1.74
1.00	1.74	1.00
1.50	2.00	.00
2.00	1.74	−1.00
2.50	1.00	−1.74
3.00	.00	−2.00
3.50	−1.00	−1.74
4.00	−1.74	−1.00
4.50	−2.00	.00
5.00	−1.74	1.00
5.50	−1.00	1.74
6.00	.00	2.00

b.

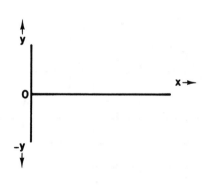

P1-20   Redraw each curve.
and P1-21

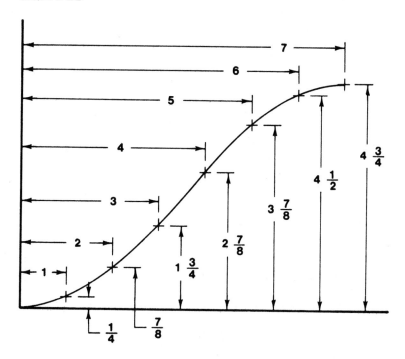

FIGURE P1-20

FIGURE P1-21

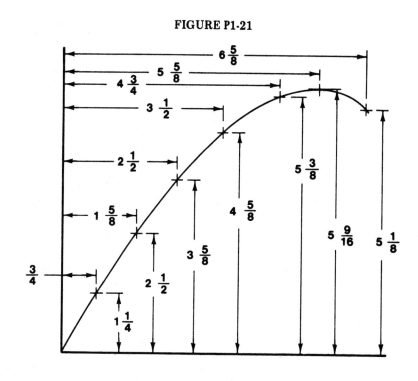

P1-22   The panel shown in Figure P1-22 is part of a monocoque
         chassis design for a dirt-track motorcycle. It was created
         by Bob Gould and Pete Morgan. Draw the panel on B-size
         (11 × 17) paper and use the 1/4 scale on your triscale.
         Label the finished drawing "Scale 1/4 = 1."

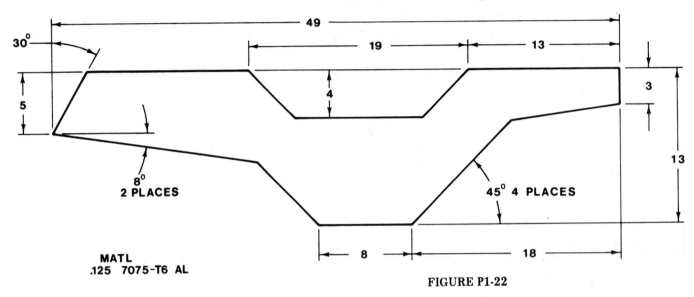

MATL
.125 7075-T6 AL

FIGURE P1-22

P1-23   The performance data for the Yamaha 350RD pictured in
         Figure P1-23 is given in the table provided. Plot the data,
         draw in the curves, label each curve, and then write a short
         paragraph explaining what the curves mean.

Engine Performance for Yamaha RD 350:

Engine Speed (rpm)	Horsepower (bhp)	Torque (ft-lb)
2000	6.0	15.0
3000	10.5	17.0
4000	15.0	19.5
5000	21.0	21.5
6000	29.0	25.0
7000	37.0	28.0
8000	36.5	23.0
9000	24.0	12.0

FIGURE P1-23   Courtesy of Yamaha Motor Corporation.

P1-24   Redraw each figure.
through
P1-30

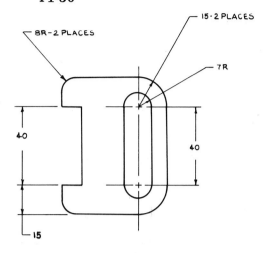

FIGURE P1-24   All dimensions are in millimeters.

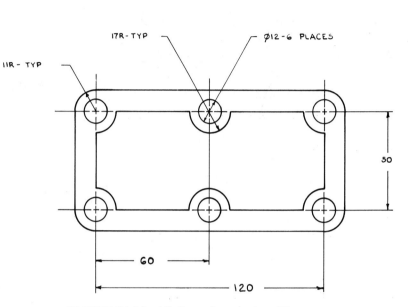

FIGURE P1-25   All dimensions are in millimeters.

FIGURE P1-26   All dimensions are in millimeters.

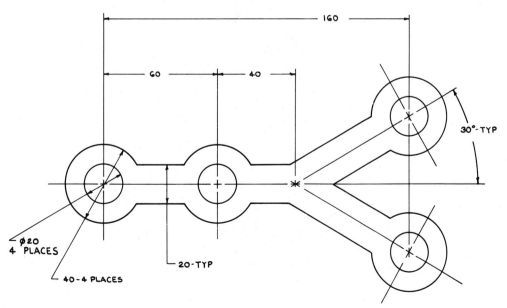

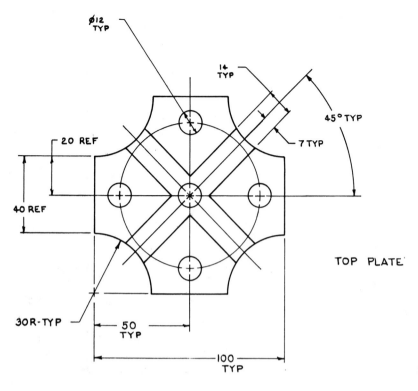

TOP PLATE

FIGURE P1-27 All dimensions are
in millimeters.

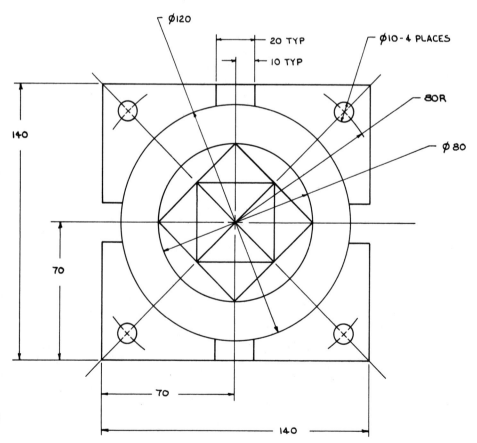

FIGURE P1-28 All dimensions are
in millimeters.

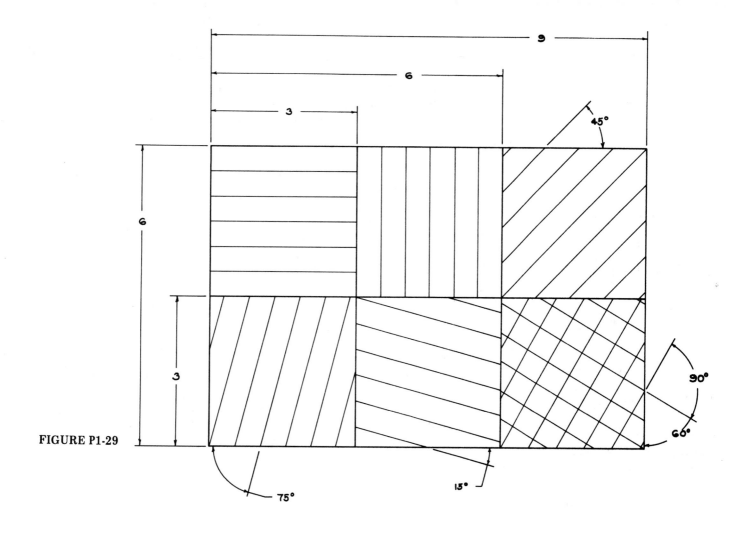

FIGURE P1-29

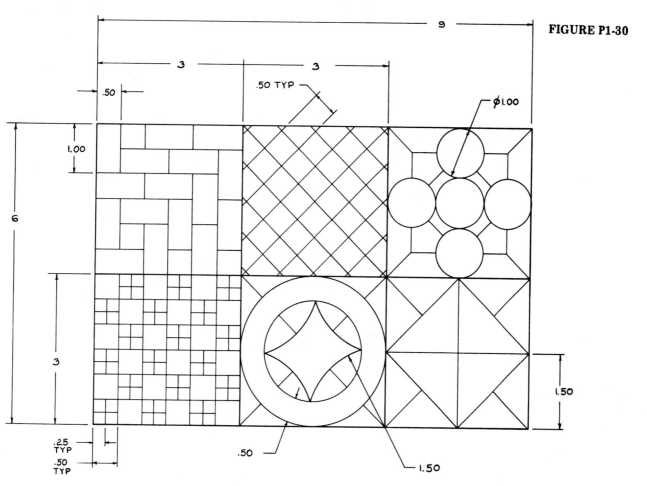

FIGURE P1-30

P1-31 Draw each shape using the dimensions given. After com-
and P1-32 pleting each figure, measure the distances called for in the
chart and add their values to the chart.

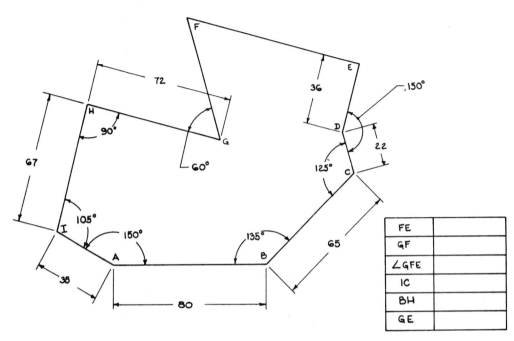

FIGURE P1-31 All dimensions
are in millimeters

FE	
GF	
∠GFE	
IC	
BH	
GE	

FIGURE P1-32

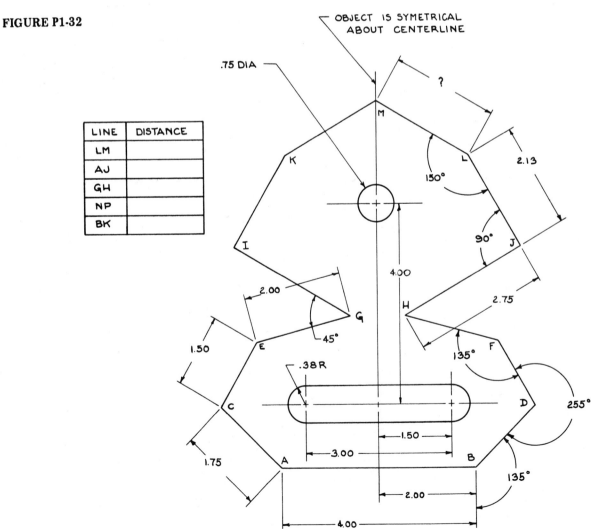

LINE	DISTANCE
LM	
AJ	
GH	
NP	
BK	

2

LINES AND LETTERS

FIGURE 2-0 Courtesy of Teledyne Post, Des Plaines, IL 60016.

2-1 INTRODUCTION

This chapter deals with drawing some of the many different kinds of lines used in technical drawings and with creating freehand lettering. Since each of these techniques will require a great deal of practice before proficiency is developed, do not be discouraged if your first attempts seem shaky. The more you draw, the better your techniques will become.

2-2 KINDS OF LINES

There are many kinds of lines commonly used on technical drawings: visible, hidden, center, leader, and phantom to name but a few. Each has its own specific configuration, thickness, intensity, and usage, some of which are defined below. They are illustrated in Figure 2-1. Those lines not defined in this chapter have a very specialized usage which will be explained in conjunction with the subjects to which they are related.

FIGURE 2-1 Five different types of lines commonly used on technical drawings.

Visible Line

Hidden Line

Leader Line

Centerline

Phantom Line

Center

Hidden

Visible

Phantom

Leader

10TS56-4

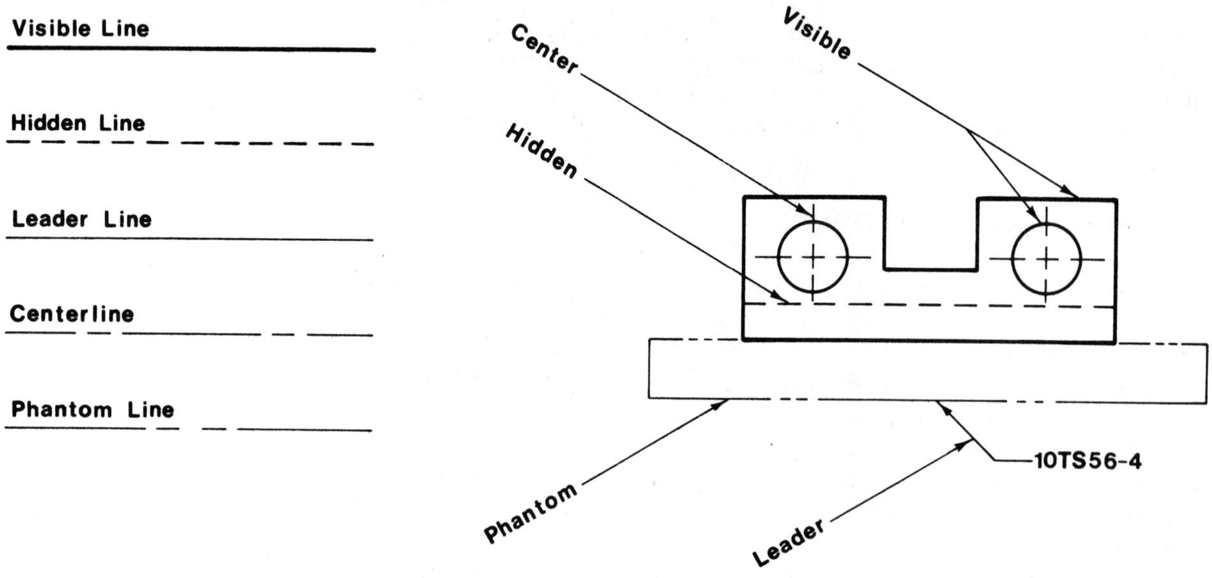

VISIBLE LINES: Heavy, thick, black lines approximately 0.020 inch thick. Uniform in color and density. It may be helpful to draw object lines with a slightly rounded lead in order to generate the necessary thickness. To form a slightly rounded lead point, first sharpen the lead and then draw a few freehand lines on a piece of scrap paper to take the initial sharpness off the lead.
—Used to define the visible edges of an object.
HIDDEN LINES: Medium, black, dashed lines approximately 0.015 inch thick. The dashes should be approximately four times as long as the intermittent spaces. Hidden lines should be a little thinner and a little lighter than visible lines.
—Used to define the edges of an object which are not directly visible. For further explanation, see Section 5-3.
LEADER LINES: Thin, black lines about 0.010 inch thick. Leader lines should be noticeably thinner (about half as thick as visible lines). To achieve the required line contrast, draw leader lines with a sharply pointed lead.
—Used to help dimension an object. For further explanation, see Section 6-2.
CENTERLINES: Thin, black lines drawn in a long-space-short line-space pattern approximately 0.010 inch thick. The long sections may be drawn at any convenient length, but the short sections must be approximately 1/8 long and the intermittent spaces should be approximately 1/16 long. Except for this configuration, centerlines are identical to leader lines.
—Used to define the center of all or part of an object. They are most commonly used to define the center of holes. They may also be used to help dimension an object.
PHANTOM LINES: Thin, black lines drawn in a long line-space-short line-space-short line configuration approximately 0.010 inch thick. The long sections may be varied in length, but the short lines must be 1/8 long and the intermittent spaces should be approximately 1/16 long.
—Used to show something that is relative to but not really part of a drawing.

After you have studied Figure 2-1, try the exercises included at the end of the chapter. Concentrate on line intensity and thickness and on the contrast between the different kinds of lines. Intensity and thickness are important, but equally important is that there be a noticeable difference between the lines. For example, visible lines must be approximately twice as thick as leader lines.

2-3 FREEHAND LETTERING

Figures 2-2 and 2-3 show the shape and style of the letters and numbers most commonly used on technical drawings. Either the vertical or inclined style is acceptable. The most widely accepted height for letters and numbers is 1/8 or 3/16, although this may vary according to the individual drawing requirements.

When you are lettering, see that the lead is tapered and slightly rounded at the tip. This differs from the tapered, sharp shape recommended for drawing lines because it is easier to draw letters and numbers with a rounded point.

Also when you letter, use a softer lead (H or 2H) because it is

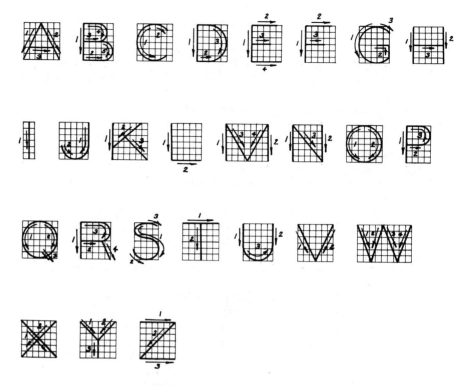

FIGURE 2-2 Vertical letters.

FIGURE 2-3 Inclined letters.

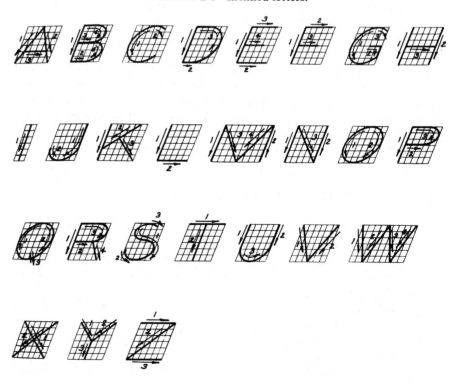

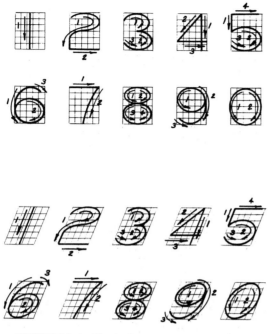

FIGURE 2-4 Vertical and inclined numbers.

easier to letter with a soft lead than with a harder one. Since soft leads tend to deposit excess amounts of graphite on the drawing, save lettering until the last phase of creating a drawing (see Figure 2-4).

2-4   GUIDE LINES

Guide lines are very light layout lines 1/8 or 3/16 apart (or whatever letter height is desired) which serve to help keep freehand lettering at a uniform height. They may be drawn with the aid of a scale and T-square or with the aid of a special guide line tool such as the Ames Lettering Guide. Drafters sometimes draw their guide lines using a nonreproducible blue pencil so that when the drawing is reproduced the guide lines seem to have disappeared and only the letters or numbers remain. Figure 2-5 illustrates guide lines.

FIGURE 2-5 Guide lines for lettering.

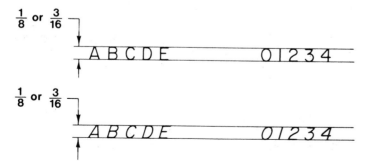

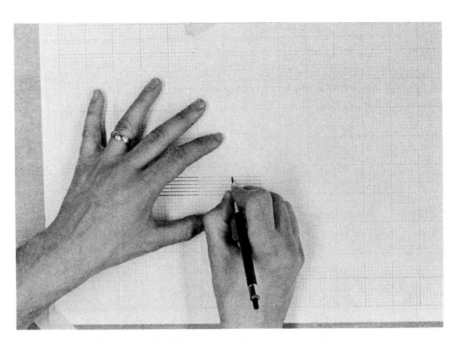

FIGURE 2-6 Creating guide lines for lettering by sliding a previously prepared set of parallel lines under the drawing paper.

Drafters sometimes avoid putting guide lines on their drawings by slipping a piece of graph paper, whose grid lines are the desired distance apart, under the paper and then lettering within the grid lines. This, of course, may be done only if you are working on a transparent medium.

If graph paper is not available or does not have the right size grid, you may make your own guide line pattern on a separate piece of paper and then slide it under your drawing as was recommended for the graph paper. Figure 2-6 shows how this is done. Save the prepared guide line pattern for future use.

2-5 LETTERING GUIDES

There are several different lettering guides that may be used to create letters and numbers for drawings. By far the most widely used guides for pencil work are the stencil kind shown in Figure 2-7.

FIGURE 2-7 Lettering guides.

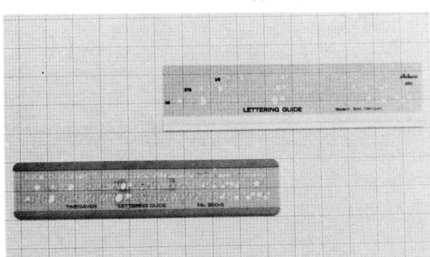

When you use a lettering guide, support it with a rigidly held T-square or other straight edge so that all the lettering is kept in the same line. Use the same lead point shape described for use in freehand lettering (see Section 2-3).

PROBLEMS

P2-1 Redraw each figure.
through
P2-5

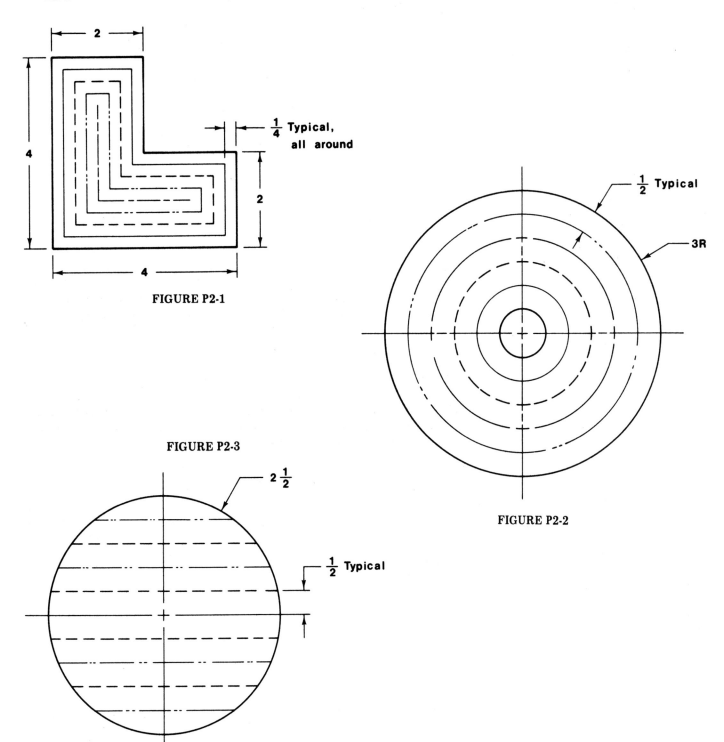

FIGURE P2-1

FIGURE P2-3

FIGURE P2-2

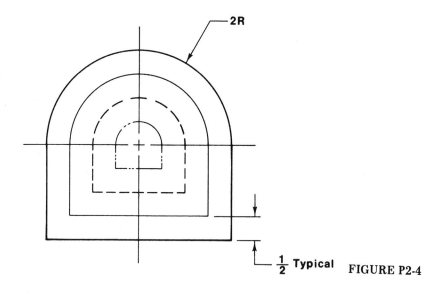

2R

$\frac{1}{2}$ **Typical**   FIGURE P2-4

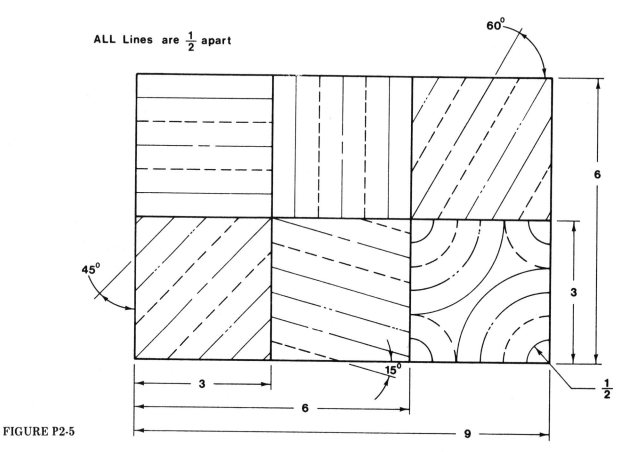

ALL Lines are $\frac{1}{2}$ apart

60°

6

45°

3

15°

3

6

9

$\frac{1}{2}$

FIGURE P2-5

P2-6   Letter the following notes.* Use the format illustrated in Figure P2-6.

1. Right shown-left-sym. opp.
2. All inside bend radii are 2-times metal thickness unless otherwise shown.

*Courtesy of Chyrsler Motors.

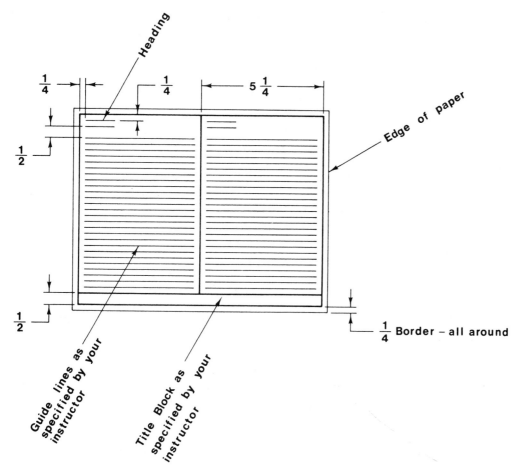

FIGURE P2-6

3. Mating surfaces must be coordinated with master die model of parts as shown.

4. Same as 3431906 except as shown & use 3799034 w/wpr motor decal in place of 3431908.

5. For inspection purposes, anchor hole at point A to set dimensions specified, and check points B & C with 50# load applied at point D in direction shown. Do not use tension spring during insp. check.

6. Open tab for access to seat track attachment. Tab must be closed after assy. of track (2-places).

7. Exceptions to PS4480-entire week may be coded with first workday of that week.

8. The terminals shall not loosen or pull off their component assy at less than 15-lb effort applied to either terminal for installation or removal of mating terminal.

9. For additional detail, see master model in Ornamentation Studio Dept 6910.

10. Vendor must obtain location approval for gating & ejector pins & dimensional approval of ejector pin bosses from materials laboratory engineering staff prior to construction.

P2-7   Using the format shown in Figure P2-6, letter in the following notes.*

1. Test per spec. P-72-B.
2. Paint black per spec. M-33.
3. Fan hood opening and fittings that fall on centerline of radiator may vary ±0.06 from centerline.
4. "Ref" dimensions are for information only and therefore will not be inspected per this print.
5. Permissible quality of hose fittings is 2% of outer diameter.
6. Finish: black paint per AM 6015 to withstand 96-hr salt spray test per AM 6015 except finish on stainless steel flexible blade to withstand 20-hr salt spray test.
7. All stamped identifications must be legible after painting.
8. Remove all fins & burrs.
9. Alternate balancing method: balance by drilling 0.33″ dia. max. holes in spider arms. The complete hole must be within 2″ of the spider o.d. & the max. depth to drill point must be 0.125″.
10. Part no. & vendor identification (C.F.-RD-69597) to be stamped in this area with 0.25″ size letters × 0.010/ 0.005 deep on one or more arms & must appear on backside of fan blade reinforcement cap.
11. Valve must be fully open (0.50 min stroke) at 8.0 ± 0.02″ HG vacuum signal on diaphragm.
12. Engineering approval of samples from each supplier is required prior to authorization of part production.

P2-8   Using the format shown in Figure P2-6, letter in the following information.†

Drafting Checking Guide

The following checklist should be used as a guide in checking drawings for compliance with related sections of Manufacturing Standard S, Drafting Standards.

1. Does the general appearance of the drawing conform to Ford Manufacturing Drafting Standards? Is the drawing clear, neat, and thorough?
2. Have the proper sheet sizes been used?
3. Has the title block been filled in completely and is the information correct? Are the title, scale, data, drawing and sheet numbers, etc., correct? Is the title complete and clear? Does the title include name of tools or equipment, operation or product part name?
4. Is the drawing number correct and according to the proper "Z" classification?

*Courtesy of American Motors.
†Courtesy of Ford Motor Company.

5. Are figures, letters, and lines correctly formed, uniform, and clean? Are they sharp and dense enough to assure good reproduction and legibility?

6. Are the necessary views and sections shown and are they positioned in proper relation to each other?

7. Do witness lines extend to the correct surface?

8. Do arrowheads extend to the correct witness lines?

9. Are all necessary dimensions shown?

10. Are drawings and dimensions to scale?

11. Are dimensions which are not to scale underlined with a wavy line, except those details with broken out sections?

12. Has duplication of dimensions and notes been avoided?

13. Are all components and included jobs shown in the stock list?

14. Has the assembly drawing been changed to agree with revised detail drawing?

15. Are related "Z" and "S" numbered tools properly listed on the main assembly drawings for reference?

16. On rework jobs are all changes fully and clearly listed?

17. Are cast details designed according to established practices? See Group XB5.

P2-9 Using the format shown in Figure P2-9, letter in the following information. Place the first line of information directly over the column headings and label each additional line *above* the preceding line.

FIGURE P2-9

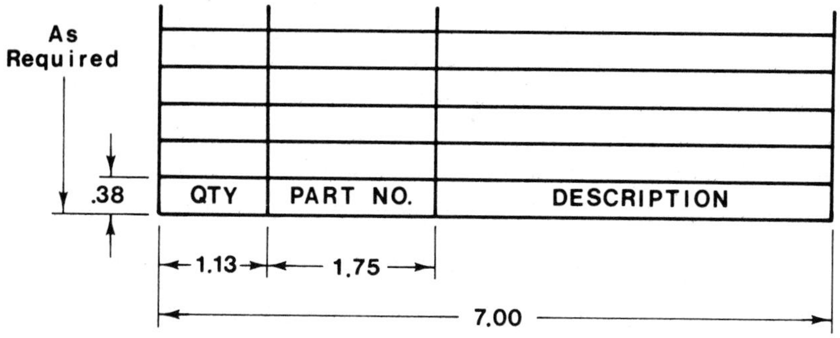

QTY.	PART NO.	DESCRIPTION
1	564S72	Housing
1	564S75	Cover Plate
4	663A46	Clips
2	100T01	Bracket
12	XX	6-32 Screws
1	564S80	Side Support R.H.
1	564S85	Side Support L.H.
1	564S90	Base Plate
4	678Q99	Dowel Pins

P2-10 Using the format shown in Figure P2-10, letter in the following information.* Use your own initials under the DR heading. Leave the CK column blank.

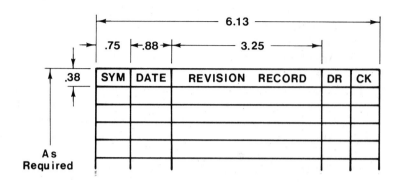

SYM	DATE	REVISION RECORD	DR	CK

SYM.	DATE	REVISION RECORD
1	7/16/69	Optional Weight Revision
2	7/16/69	1.742/1.729 was 1.747/1.734
3	7/16/69	Note 8 Relocated
	8/11/69	Issued
	9/18/69	Released
A	5/12/70	See SF3210686 Rev A
B	10/21/70	Surface "J" Added
C	10/21/70	Notes 8 & 9 Added
D	10/21/70	Note Added
E	11/17/71	4 was 3
F	3/2/72	1.743/1.728 was 1.742/1.729
G	3/28/72	Was .005 T.I.R.

FIGURE P2-10

P2-11 Prepare a drawing using the format shown in Figure P2-11. Complete the drawing by repeating each letter until the next consecutive letter is reached. Continue lettering until all letters in the alphabet have been drawn. Keep all guide lines very thin and light.

*Courtesy of American Motors.

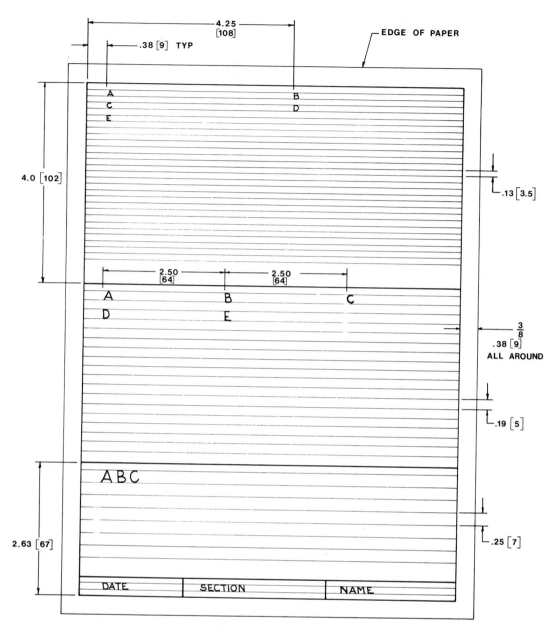

FIGURE P2-11

P2-12 Repeat Problem P2-11 using inclined letters.

P2-13 Prepare a drawing using the format shown in Figure P2-13. Complete the drawing by repeating each number until each line is completely filled. Keep all guide lines very thin and light. Add your name; date, and section in the three blocks at the bottom of the drawing. Figure P2-13 uses the same format as Figure P2-11.

P2-14 Repeat Problem P2-13 using inclined letters.

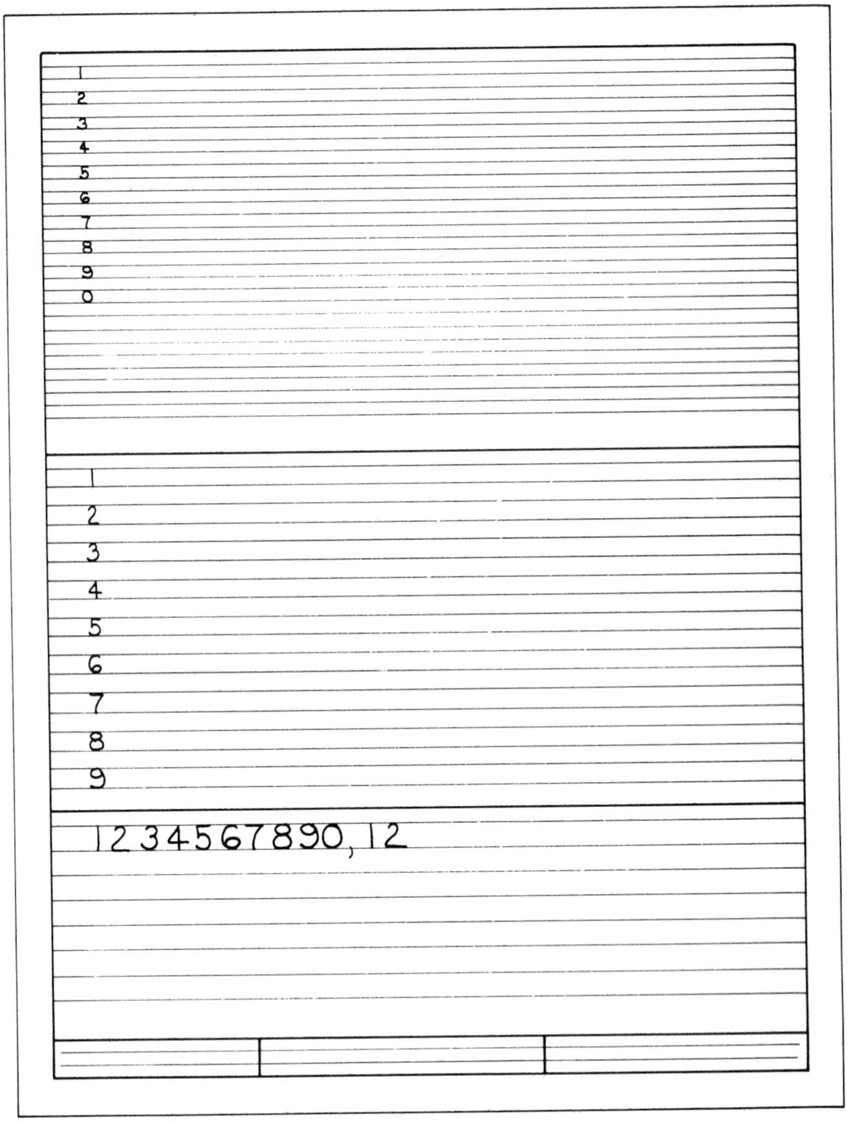

FIGURE P2-13

P2-15   Prepare a table listing the equivalents between fractions and decimal inches. Use the format shown in Figure P2-15. Use values for every 1/64 of an inch.

P2-16   Prepare a table listing the equivalents between fractional inches and millimeters. Use the format shown in Figure P2-15. Figure P2-16 shows a sample of how the finished table should appear.

P2-17   Prepare a table listing the equivalents between millimeters and decimal inches. Use the format shown in Figure P2-15. Figure P2-17 shows a sample of how the finished table should appear.

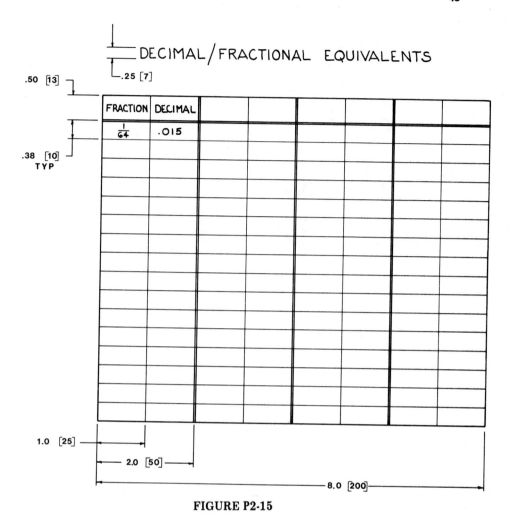

FIGURE P2-15

FIGURE P2-16

INCH	mm	INCH	
$\frac{1}{64}$	.397		
$\frac{1}{32}$	.794		
$\frac{3}{64}$	1.190		
$\frac{1}{8}$			

FIGURE P2-17

mm	INCH	mm	
1	.039		
2	.079		
3	.118		
4			

3

GEOMETRIC
CONSTRUCTIONS

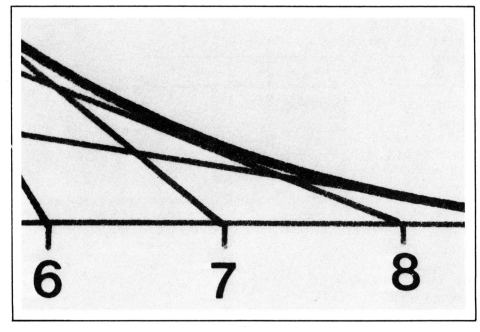

FIGURE 3-0

3-1 INTRODUCTION

Geometric constructions are the building blocks of drafting. Every drawing, regardless of its difficulty, is a composite of geometric shapes. A rectangle is four straight lines and four right angles. A cam is a series of interconnected arcs of various radii. Every drafter must have a fundamental knowledge of geometric constructions if he or she is to progress to the more difficult format and layout concepts required by most drawings.

This chapter is set up for easy reference. Each page contains one method of doing one geometric construction. Both classical methods and those requiring drafting equipment are presented. No attempt has been made to avoid redundancy, and each method is completely described within the page on which it is presented. A list of all constructions described in this chapter is given below.

3-2   POINTS AND LINES

A point, to a drafter, is defined by the intersection of two construction lines.

Note:

A dot should not be used to define a point because a dot may be easily confused with other marks on the drawing and thereby cause errors.

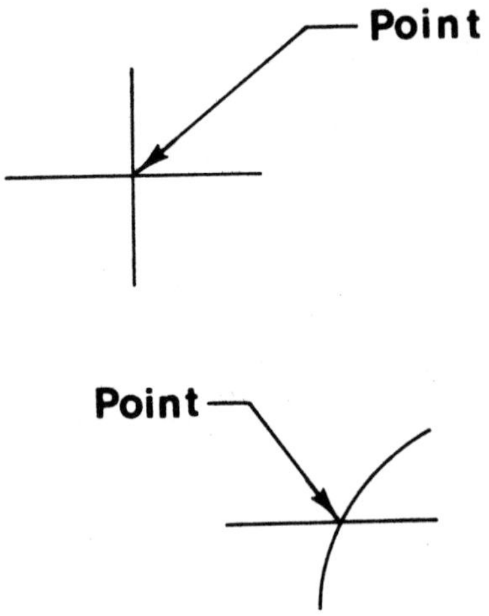

Point

Point

A line, to a drafter, is an object line connecting two or more points.

Note:

The accuracy of a curved line depends on the number points used to define it. The number of points used depends on the accuracy required for the particular curve.

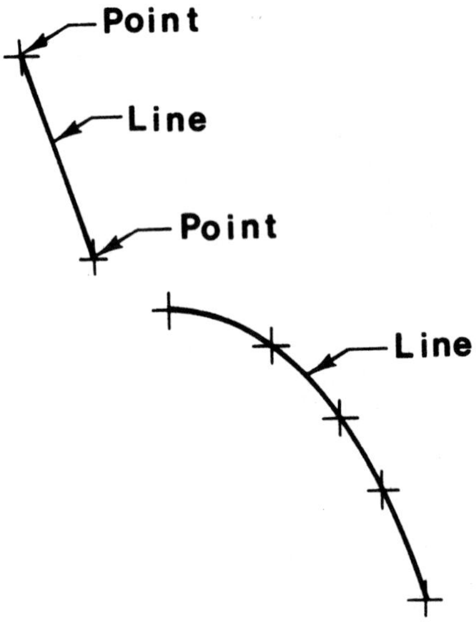

Point

Line

Point

Line

3-3   ADD AND SUBTRACT LINES

Given: Line 1-2 of length X and line 3-4 of length R.

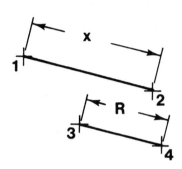

Problem: Add line 1-2 to 3-4.
1. Construct a line and define point 1 anywhere along it.

2. Using a compass set on point 1, construct an arc of radius X.
3. Using a compass set on the intersection of the arc constructed in step 2 and the line constructed in step 1, construct an arc of radius R as shown. Line 1-4 is equal to line 1-2 plus line 3-4.

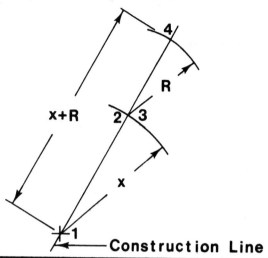

Given: Line 1-2 of length X and line 3-4 of length R.

Problem: Subtract line 3-4 from line 1-2.

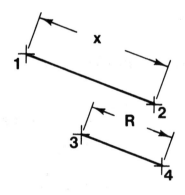

1. Construct a line and define point 1 anywhere along it.
2. Using a compass set on point 1, construct an arc of radius X.

3. Using a compass set on the intersection of the arc constructed in step 2 and the line constructed in step 1, construct an arc of radius R as shown.

Line 1-4 equals line 1-2 minus line 3-4.

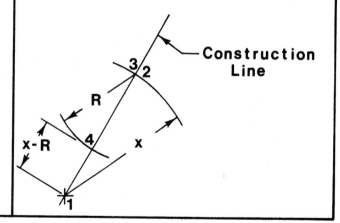

3-4   PARALLEL LINES — FIRST METHOD

Given: Line 1–2 and distance D.

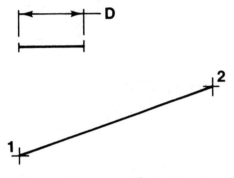

Problem: Construct a line parallel to line 1–2 at distance D.

1. Using a compass set anywhere along line 1–2, construct an arc of radius D as shown.

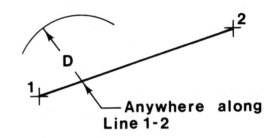

Anywhere along Line 1-2

2. Construct another arc of radius D as shown.

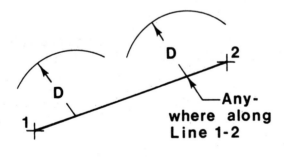

Any- where along Line 1-2

3. Construct a line tangent to both arcs.

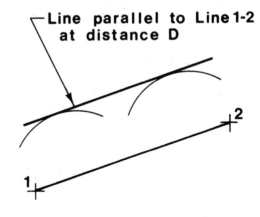

Line parallel to Line 1-2 at distance D

3-5   PARALLEL LINES — TWO-TRIANGLE METHOD

Given: Line 1–2 and distance D.

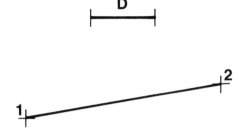

Problem: Construct a line parallel to line 1–2 at a distance D.

1. Align the shortest leg of a 30–60–90 triangle with line 1–2.

2. Place the hypothenuse of a 45–45–90 triangle against the hypothenuse of the 30–60–90 triangle.

3. Holding the 45–45–90 triangle firmly and in place, slide the 30–60–90 along the hypothenuse of the 45–45–90 as shown.

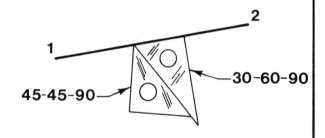

4. Construct a line along the edge of the 30–60–90 as shown.

5. Mark off a distance D from line 1–2 along the line constructed in step 4.

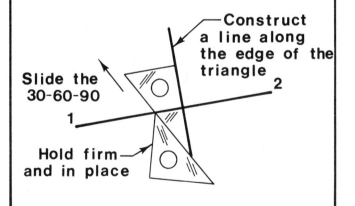

6. Realign the triangles to line 1–2 and slide the 30–60–90 until it is a distance D from line 1–2.

7. Construct a line along the shortest leg of the 30–60–90 through distance D parallel to line 1–2.

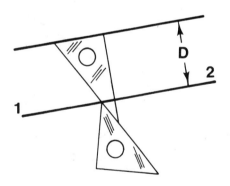

Note: ° A T-square may be used in lieu of a second triangle.
  ° The line used to locate distance D from line 1–2 in step 5 is perpendicular to line 1–2.

3-6   BISECT A LINE — FIRST METHOD

Given: Line 1-2.

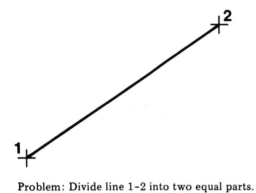

Problem: Divide line 1-2 into two equal parts.

1. Construct an arc of radius R. Use point 1 as center.

R = any radius of greater length than ½ line 1-2.

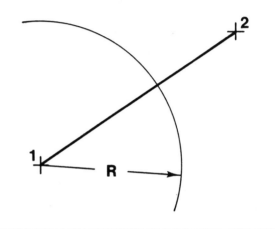

2. Construct an arc of radius R. Use point 2 as center.

3. Define the intersection of the arcs as points 3 and 4.

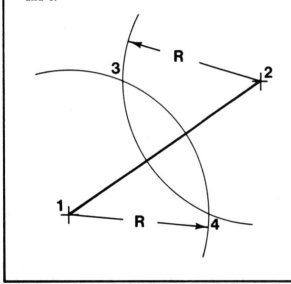

4. Connect points 3 and 4 with a construction line.

5. Define point 5 where line 3-4 intersects line 1-2. Line 1-5 = line 5-2.

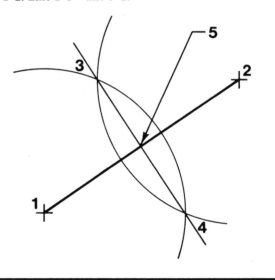

Note: This is the classical method as taught in plane geometry.

3-7   BISECT A LINE — SECOND METHOD

Given: Line 1-2.

1 ├─────────────────────────┤ **2**

Problem: Divide line 1-2 into two equal parts.

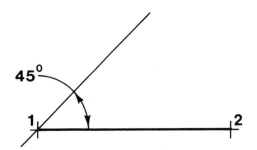

1. Align the T-square with line 1-2 and using a 45-45-90 triangle as a guide, construct a line 45° to line 1-2 through point 1.

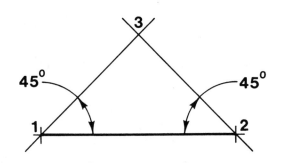

2. Repeat step 1 this time constructing the 45° line through point 2.

3. Define the intersection of the construction lines as point 3.

4. Draw a line through point 3 perpendicular to line 1-2 which intersects line 1-2.

5. Define point 4 as shown. Line 1-4 = line 4-2.

Note: This method relies on drafting equipment for completion. Any angle may be used in steps 1 and 2 as long as they are equal.

3-8   DIVIDE A LINE INTO ANY NUMBER
OF EQUAL PARTS

Given: Line 1-2.

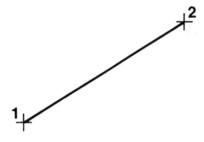

Problem: Divide line 1-2 into five equal parts.

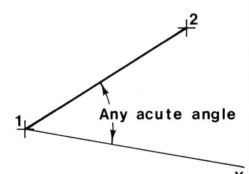

1. Construct a line A-X at any acute angle to line 1-2.

2. Mark off five equal spaces along line 1-X and construct a line 2-7.

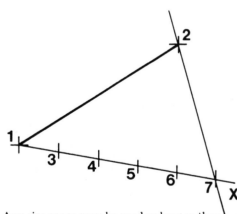

Note: Any size space may be used as long as they are all equal in length.

3. Draw lines 6-F, 5-E, 4-D, and 3-C parallel to line 2-7.

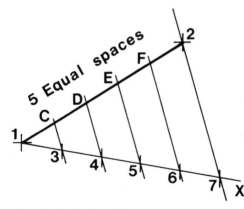

1-C = C-D = D-E = E-F = F-2.

Note: This method is good for any number of equal parts, not just for the five shown. Once 1-X has been drawn, mark off as many spaces as needed. Remember that the spaces must be of equal length.

3-9   DIVIDE A LINE INTO PROPORTIONAL PARTS

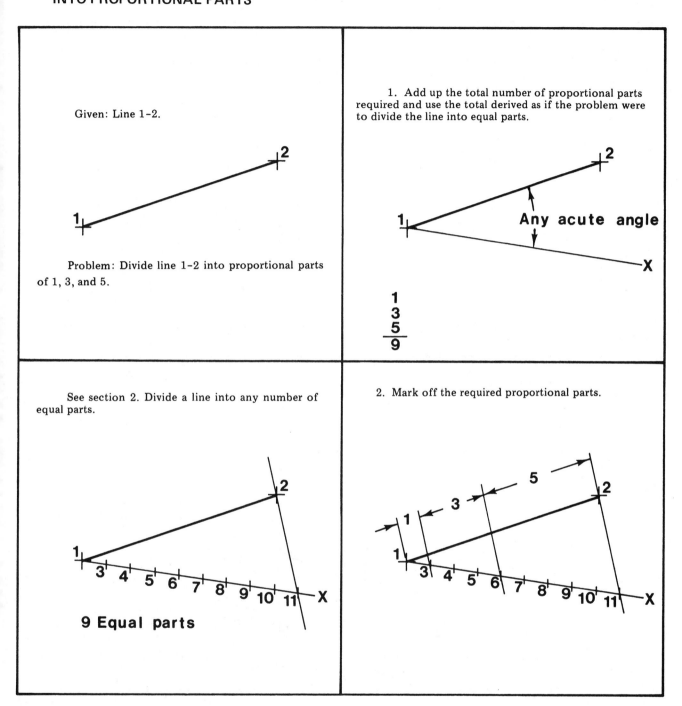

Given: Line 1-2.

Problem: Divide line 1-2 into proportional parts of 1, 3, and 5.

1. Add up the total number of proportional parts required and use the total derived as if the problem were to divide the line into equal parts.

Any acute angle

$$\begin{array}{r} 1 \\ 3 \\ \underline{5} \\ 9 \end{array}$$

See section 2. Divide a line into any number of equal parts.

9 Equal parts

2. Mark off the required proportional parts.

Note: This method is good for any number of parts and any ratio, not just for the 1, 3, and 5 shown.

3-10   FILLETS — RIGHT ANGLES ONLY

Given: Right angle and radius R.

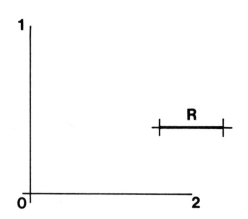

Problem: Draw a fillet of radius R tangent to angle 1-0-2.

1. Construct an arc of radius R. Use point 0 as center.

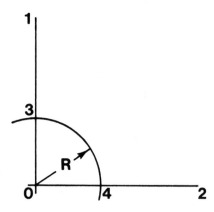

2. Define points 3 and 4 where the arc intersects lines 0-1 and 0-2.

3. Construct two more arcs of radius R. Use points 3 and 4 as centers.

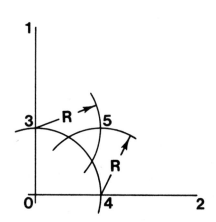

4. Define point 5 where the arcs centered at points 3 and 4 intersect.

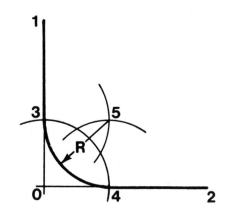

5. Draw a fillet of radius R. Use point 5 as center, tangent to lines 0-1 and 0-2.

Note: • This method is good *only* for right angles.
   • For small radii, use a circle template and draw fillet directly.

3-11   FILLETS — ANY ANGLE

Given: Angle 2-1-3 and radius R.

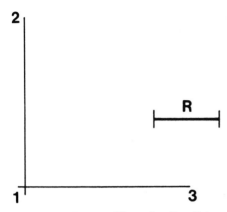

Problem: Draw a fillet of radius R tangent to angle 2-1-3.

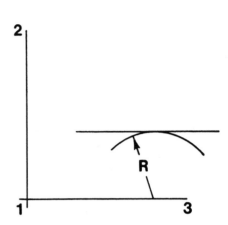

1. Construct a line parallel to line 1-3 at a distance R.

2. Construct a line parallel to line 1-2 at a distance R.

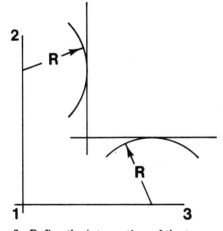

3. Define the intersection of the two constructed parallel lines as point 4.

4. Draw a fillet of radius R tangent to angle 2-1-3.

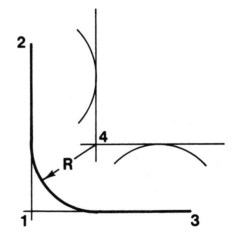

Note: • This method is good, not only for the right angle as shown, but also for any angle, acute or obtuse.
 • For small radii, use a circle template and draw the fillet directly.

3-12 ROUNDS — ANY ANGLE

Given: Angle 1-0-2 and radius R.

Problem: Draw a round of radius R tangent to angle 1-0-2.

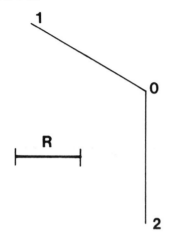

1. Construct a line parallel to line 0-2 at a distance R.

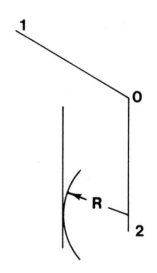

2. Construct a line parallel to line 0-1 at a distance R.

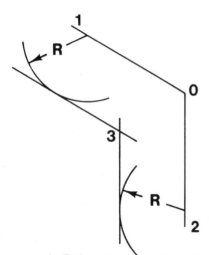

3. Define the intersection of the two constructed parallel lines as point 3.

4. Draw a round of radius R. Use point 3 as center, tangent to angle 1-0-2.

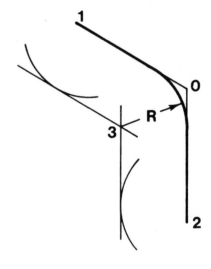

3-13   HEXAGON — FIRST METHOD

Problem: Draw a hexagon D across the corners.

1. Construct a circle of diameter D.

Note: $\dfrac{\text{Diameter}}{2}$ = radius. Set compass to radius dimension.

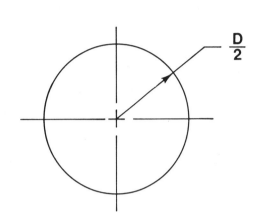

2. Using a compass, mark off six distances $\dfrac{D}{2}$ as shown.

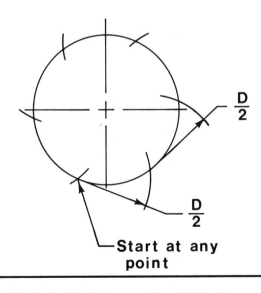

Start at any point

3. Draw in the hexagon.

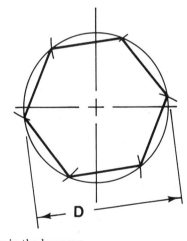

Note: This is the classical geometric method and is not generally used by drafter because it makes positioning of the hexagon difficult.

3-14   HEXAGON — SECOND METHOD

Problem: Draw a hexagon S across the corners.

1. Construct a circle of diameter S.

Note: $\dfrac{\text{Diameter}}{2}$ = radius. Set compass to radius dimension.

2. Define points 2 and 3 as shown.

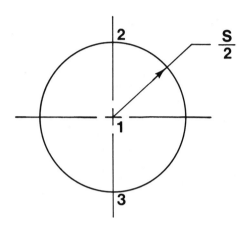

3. Using points 2 and 3 as center, construct two arcs of radius $S/2$.

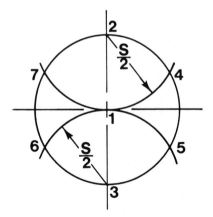

4. Define points 4, 5, 6, and 7 as shown.

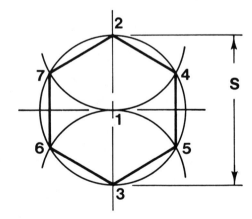

5. Draw in the hexagon.

3-15   HEXAGON — THIRD METHOD

Problem: Construct a hexagon A across the corners.

1. Construct a circle of diameter A.

2. Using a 60° triangle, construct lines 60° to the horizontal as shown.

3. Define points 1, 2, 3, and 4.

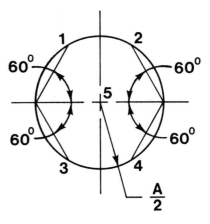

4. Construct lines 1–2 and 3–4.

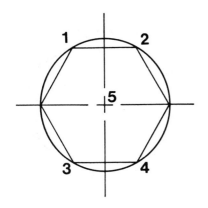

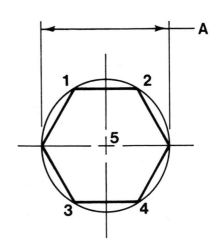

5. Draw in the hexagon.

3-16   HEXAGON – FOURTH METHOD

Problem: Construct a hexagon B across the corners.

1. Construct a circle of diameter B.

2. Using a 30° triangle, construct lines 30° to the horizontal as shown.

3. Define points 1, 2, 3, and 4.

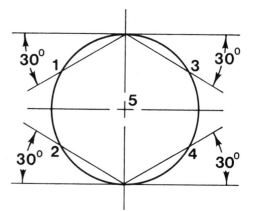

4. Construct lines 1–2 and 3–4.

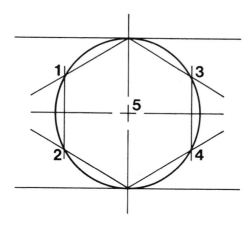

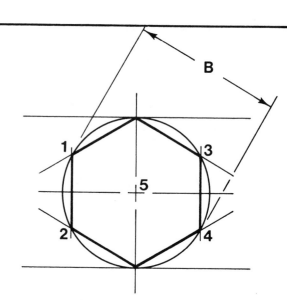

5. Draw in the hexagon.

3-17   HEXAGON — FIFTH METHOD

Problem: Construct a hexagon C across the flats.

1. Construct a circle of diameter C.

2. Construct two vertical lines tangent to the circle.

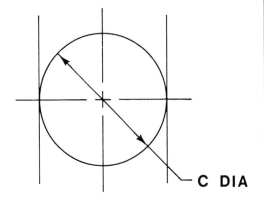

C DIA

3. Using a 30° triangle, construct lines tangent to the circle 30° to the horizontal as shown.

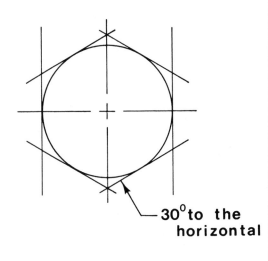

30° to the horizontal

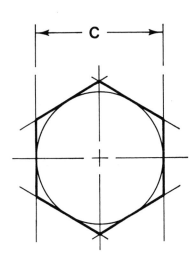

C

4. Draw in the hexagon.

3-18   PENTAGON – HOW TO DRAW

Problem: Draw a pentagon inscribed in a circle of diameter A.

1. Construct a circle of diameter A.
2. Define points 0, 1, and 2 as shown.
3. Bisect line 0–1 and define the midpoint as point 3.
4. Define point 4 as shown.
5. Using a compass set on point 3, construct an arc through point 4 and line 2–0.
6. Define the intersection of the arc constructed in step 5 and line 2–0 as point 5.

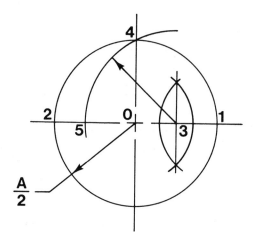

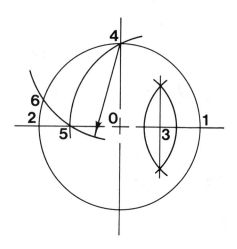

7. Using a compass set on point 4, construct an arc through point 5 and the edge of the circle.

8. Define the intersection of the arc constructed in step 7 and the circle as point 6.

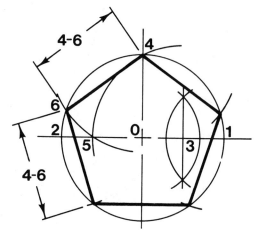

9. Using a compass, mark off the distance 4–6 around the circumference of the circle as shown.

10. Draw in the pentagon.

3-19   PENTAGON — DEFINITION

Define a pentagon.

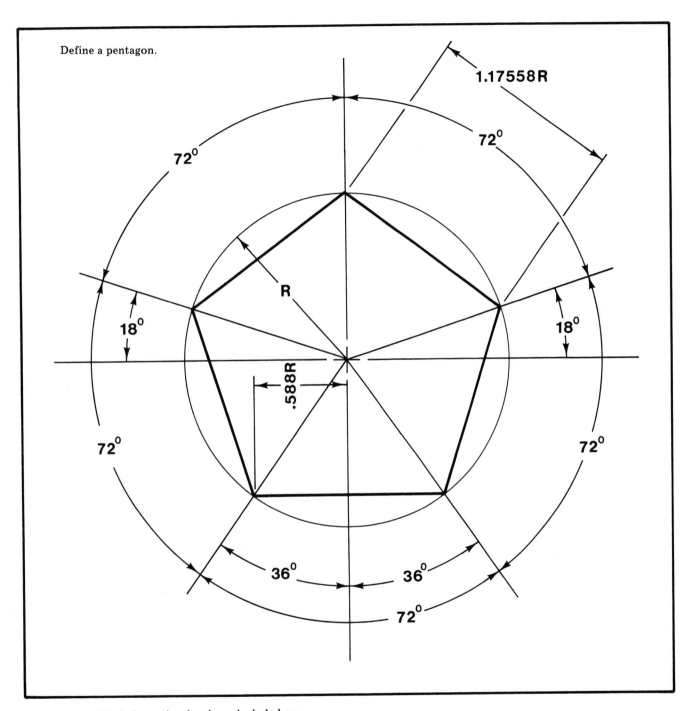

Note: This information has been included as a
reference to help in drawing pentagons.

3-20   OCTAGON

Problem: Draw an octagon D across the flats.

1. Draw a circle of diameter D.

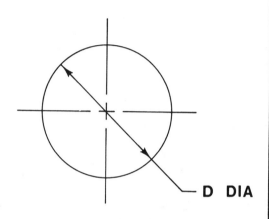

D DIA

2. Construct four tangent lines as shown.

3. Construct four lines, 45° to the horizontal, tangent to the circle as shown.

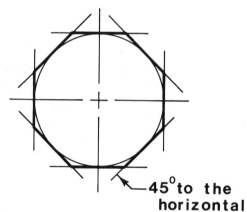

45°to the horizontal

4. Draw in the octagon.

3-21   FILLET — TWO CIRCLES

Given: Circles X and Y and radius R.

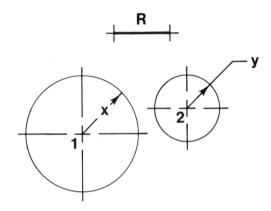

Problem: Draw a fillet of radius R tangent to circles X and Y.

1. Construct an arc of radius $X + R$. Use point 1 as center.

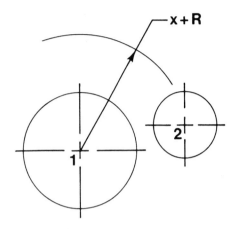

(See Section 3-3, Add and Subtract Lines.)

2. Construct an arc of radius $Y + R$. Use point 2 as center.

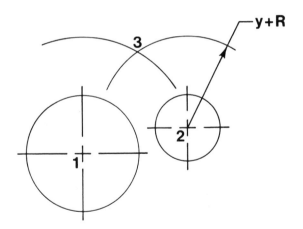

3. Define the intersection of the two arcs as point 3.

4. Using point 3 as center, draw a fillet of radius R tangent to the two circles.

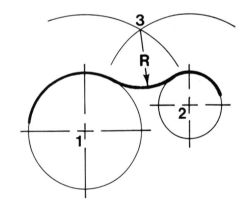

3-22 ROUND — TWO CIRCLES

Given: Circles *x* and *y* and radius *R*.

Problem: Draw a round of radius *R* tangent to circles *x* and *y*.

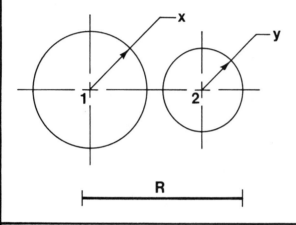

1. Construct an arc of radius $R\text{-}x$. Use point 1 as center.

(See Section 3-3, Add and Subtract Lines.)

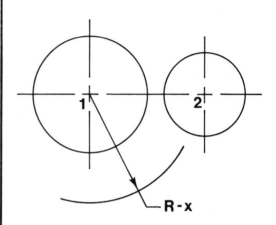

2. Construct an arc of radius $R\text{-}y$. Use point 2 as center.

3. Define the intersection of the two arcs as point 3.

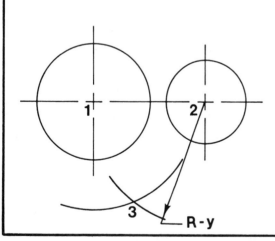

4. Using point 3 as center, draw an arc of radius *R* tangent to the two circles.

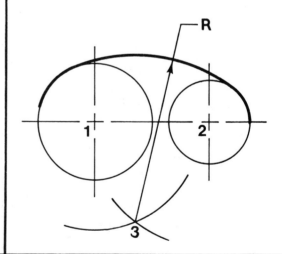

3-23   FILLET — CONCAVE CIRCLE TO A LINE

Given: A circle of radius X, line 1–2, and a radius R.

Problem: Draw a fillet of radius R tangent to a circle and a line.

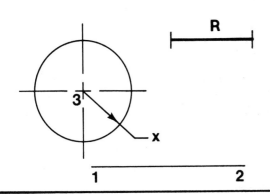

1. Construct an arc of radius $X + R$. Use point 1 as center.

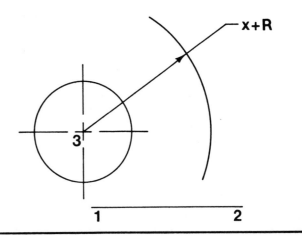

2. Construct a line parallel to line 1–2 at a distance R.

3. Define the intersection of the arc $(X + R)$ and the line parallel to line 1–2 as point 4.

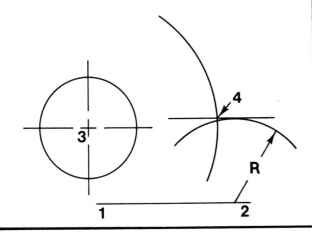

4. Using point 4 as center, draw in a fillet of radius R.

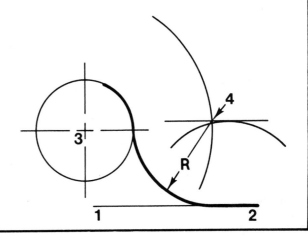

3-24   ROUND — CONVEX CIRCLE
TO A LINE

Given: Circle x, line 1–2, and fillet radius R.

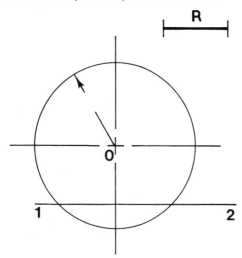

Problem: Draw a round tangent to a circle of radius x and line 1–2.

1. Construct a circle of radius $x - R$.

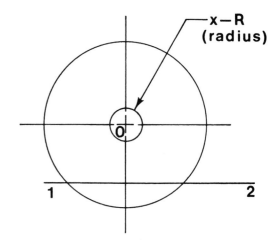

(See Section 3-3, Subtracting Lines)

2. Construct a line parallel to line 1–2 at a distance R.

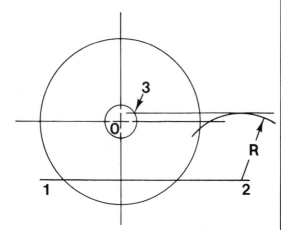

3. Define the intersection of the circle $(x - R)$ and the line parallel to line 1–2 as point 3.

4. Draw a round of radius R. Use point 3 as center.

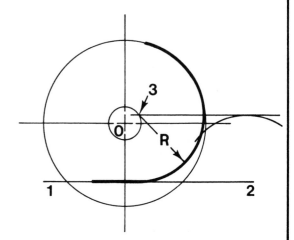

3-25   S-CURVE (REVERSE OR OGEE CURVE)

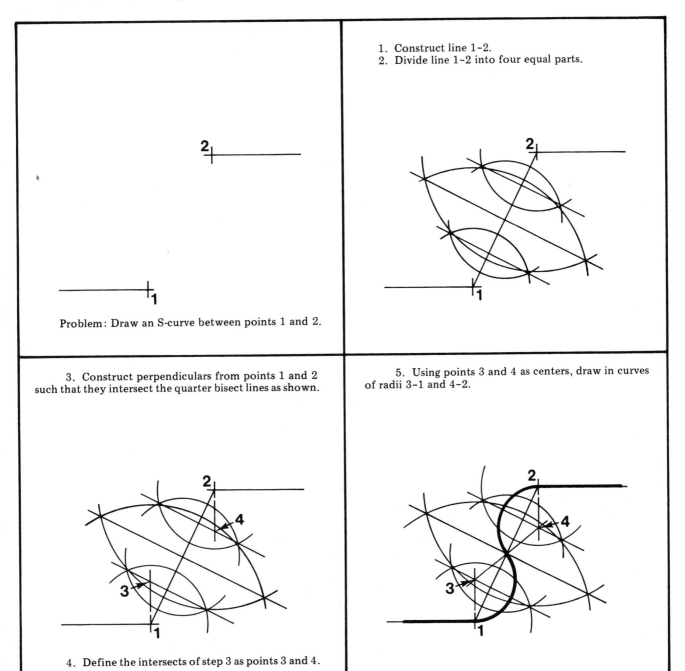

Problem: Draw an S-curve between points 1 and 2.

1. Construct line 1–2.
2. Divide line 1–2 into four equal parts.

3. Construct perpendiculars from points 1 and 2 such that they intersect the quarter bisect lines as shown.

4. Define the intersects of step 3 as points 3 and 4.

5. Using points 3 and 4 as centers, draw in curves of radii 3–1 and 4–2.

Note: The S-curve need not be symmetrical. Asymmetrical curves may be constructed, but the method is not covered in this book.

3-26   APPROXIMATE ELLIPSE

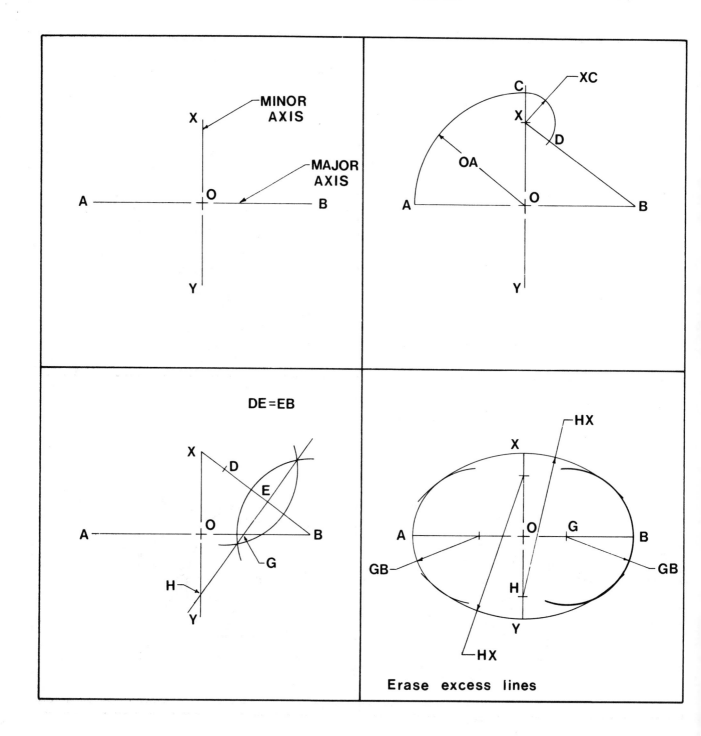

3-27   BISECT AN ANGLE

Given: Angle 1-0-2

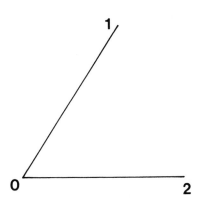

Problem: Bisect angle 1-0-2.

1. Construct an arc of radius R. Use point 0 as center.

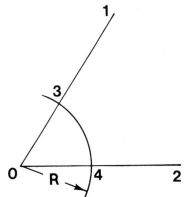

3. Using points 3 and 4 as centers, construct two more arcs of radius R as shown.

2. Define points 3 and 4 where the arc intersects lines 0-1 and 0-2.

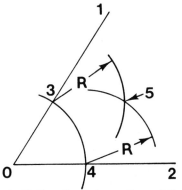

4. Define the intersection of the two arcs as point 5.

5. Construct a line 0-5.

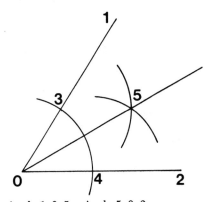

Angle 1-0-5 = Angle 5-0-2.

3-28 PARABOLA

Problem: Draw a parabola whose major axis is twice the minor axis.

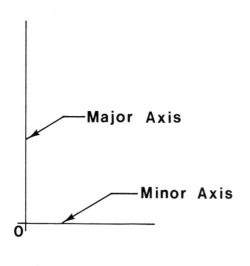

1. Lay out points making those on the major axis twice as far apart as those on the minor axis.

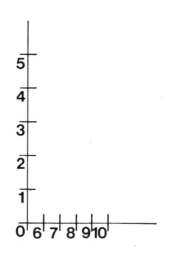

The number of points on each axis must be equal.

2. Construct lines from the first point on the major axis to the last point on the minor axis, etc. (1–10, 2–9, 3–8, 4–7, 5–6)

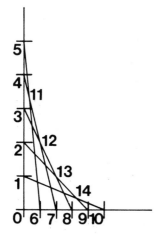

3. Define points 11, 12, 13, and 14 as shown.

4. Draw a parabola by using a French curve and connecting points 5–11–12–13–14–10.

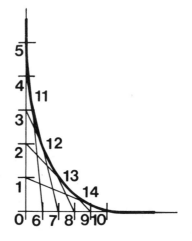

Note: °In this example the major axis points are ½ apart and the minor axis points are ¼ apart.
°The accuracy of the parabola depends on the number of points used to define it. The more points, the greater the accuracy.

PROBLEMS

Problems 3-1 through 3-41 are geometric constructions. Place four constructions per 8½ × 11 sheet using the format shown in Figure P3-1.

P3-1   You are given a line 1-15/16 long. Bisect it.

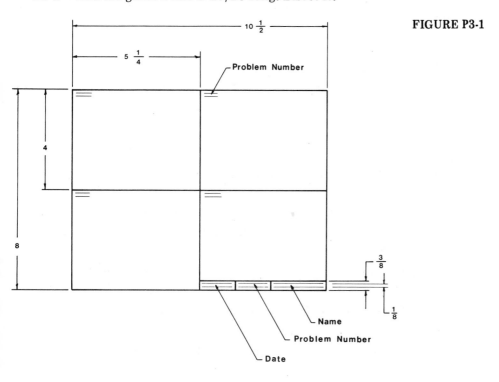

FIGURE P3-1

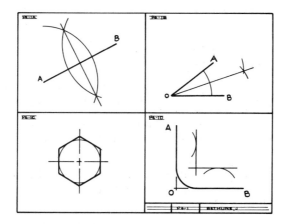

P3-2   You are given a line 2.36 long. Bisect it.

P3-3   Draw two parallel lines 15/16 apart.

P3-4   Draw two parallel lines 1-3/4 apart.

P3-5   Draw two parallel lines 2.063 apart.

P3-6   You are given a line 2-1/8 long. Divide it into 7 equal parts.

P3-7   You are given a line 3-1/4 long. Divide it into 15 equal parts.

P3-8   You are given a line 1.68 long. Divide it into 3 equal parts.

P3-9   You are given a line 1-7/8 long. Divide it into proportional parts of 2, 4, and 7.

P3-10   You are given a line 2-9/16 long. Divide it into proportional parts of 1, 4, 3, and 5.

P3-11   You are given a line 2.78 long. Divide it into proportional parts of 3, 4, and 9.

P3-12   You are given lines of 1-1/16 and 5/8 long. Graphically add them.

P3-13   You are given lines of 1.75 and 0.625 long. Graphically add them.

P3-14   You are given a line 2-3/8 long. Subtract a line 1-3/16 long from it.

P3-15   You are given a line 1-7/8 long. Subtract a line 15/16 long from it.

P3-16   You are given a line 1.28 long. Subtract a line 0.80 long from it.

P3-17   You are given a 37° angle. Draw a fillet of radius 3/4.

P3-18   You are given a 123° angle. Draw a fillet of radius 1-3/8.

P3-19   You are given a 90° angle. Draw a round of radius 7/8.

P3-20   You are given a 60° angle. Draw a round of radius 1.20.

P3-21   Draw a hexagon 1-3/4 across the corners.

P3-22   Draw a hexagon 2-3/16 across the corners.

P3-23   Draw a hexagon 1.80 across the corners.

P3-24   Draw a hexagon 80 mm across the corners.

P3-25   Draw a hexagon 2 across the flats.

P3-26   Draw a hexagon 2-9/16 across the flats.

P3-27   Draw a pentagon inscribed within a 2-3/4 diameter circle.

P3-28   Draw an octagon inscribed within a 1-7/8 diameter circle.

P3-29   You are given two circles of 1-1/2 and 7/8 in diameter and located 1-1/2 apart. Draw a fillet between them of radius 3/4.

P3-30   You are given two circles of 1.75 and 1.10 in diameter and located 1.60 apart. Draw a fillet between them of radius 0.90.

P3-31   You are given two circles of 1-1/4 and 1-1/2 in diameter and located 2-1/4 apart. Draw a round between them of radius 2-1/2.

P3-32   You are given a circle 1-7/16 in diameter located 2-1/8 above a line. Draw a fillet between them of radius 1-1/16.

P3-33   You are given a circle 2-1/4 in diameter located 9/16 above a line. Draw a round between them of radius 7/8.

P3-34   Draw an approximate ellipse with a minor axis of 1 and a major axis of 2.

P3-35   Draw an approximate ellipse with a minor axis of 2.25 and a major axis of 3.80.

P3-36   Draw a parabola whose major axis is 1-1/2 times the minor axis.

P3-37   Draw a parabola whose major axis is two times the minor axis.

P3-38   You are given an angle of 60°. Bisect it.

P3-39   You are given an angle of 50°. Bisect it.

P3-40   You are given an angle of 108°. Bisect it.

P3-41   You are given an angle of 42.5°. Bisect it.

P3-42   Redraw each shape.
through
P3-53

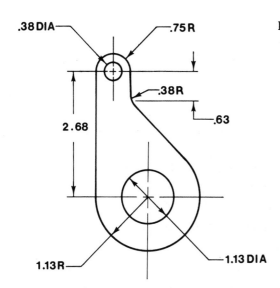

FIGURE P3-42

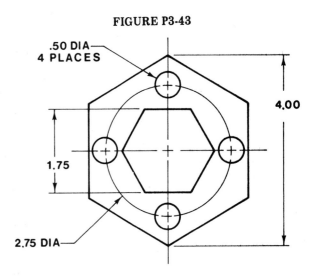

FIGURE P3-43

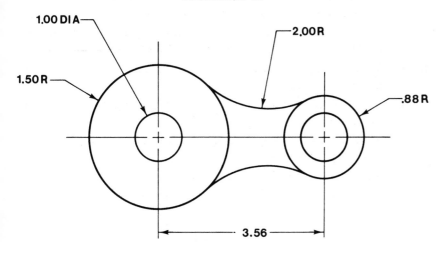

FIGURE P3-44

FIGURE P3-45

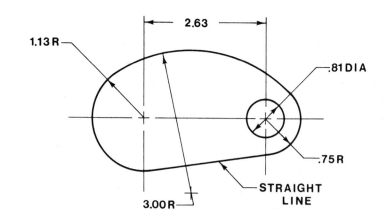

FIGURE P3-46   Use B-size paper.

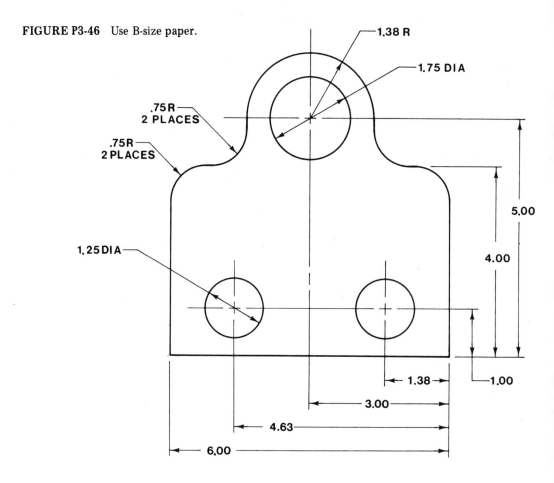

FIGURE P3-47

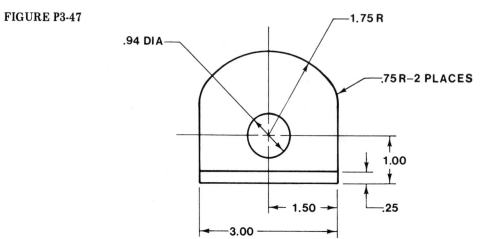

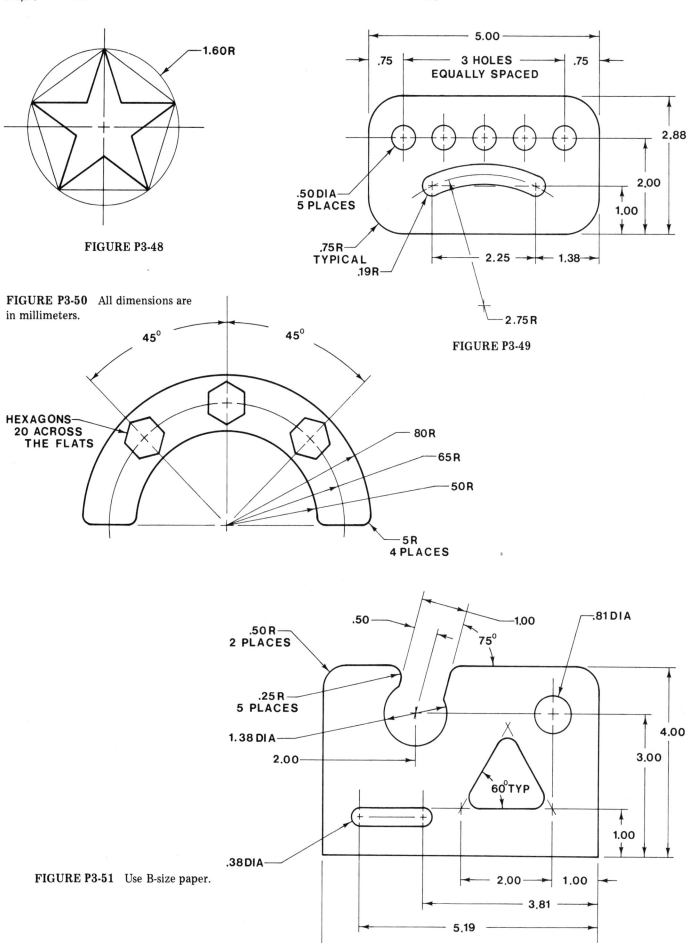

1.60R

FIGURE P3-48

5.00

.75   3 HOLES
EQUALLY SPACED   .75

2.88

2.00

1.00

.50DIA
5 PLACES

.75R
TYPICAL

.19R

2.25   1.38

2.75R

FIGURE P3-49

FIGURE P3-50   All dimensions are in millimeters.

45°   45°

HEXAGONS
20 ACROSS
THE FLATS

80R

65R

50R

5R
4 PLACES

.50R
2 PLACES

.50   1.00

75°

.81DIA

.25R
5 PLACES

1.38 DIA

2.00

60°TYP

4.00

3.00

1.00

.38DIA

2.00   1.00

3.81

5.19

6.00

FIGURE P3-51   Use B-size paper.

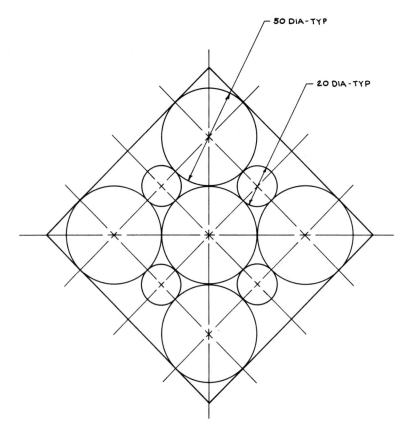

FIGURE P3-52   All dimensions are in millimeters.

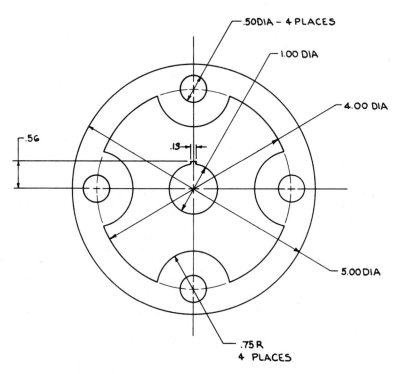

FIGURE P3-53

PROJECTION THEORY

FIGURE 4-0 Many views of an object are needed to present a clear idea of the object's shape. (Courtesy of General Motors Corp.)

4-1 INTRODUCTION

The purpose of a technical drawing is to communicate information. As in any kind of communication, it is easy to know what you want to say but sometimes very difficult to make yourself understood.

From the time a customer places the initial order until the finished product is delivered, many people of varying technical skills and backgrounds will contribute to help satisfy the demands of the order. An engineer and drafter design the product and prepare the necessary drawings. The drawings are then used progressively by a planner to price and time the job, by a buyer to order necessary manufacturing stock, by a shop supervisor to schedule and assign the work, by a machinist to actually make the parts, by an inspector to make sure that the work has been done properly, by an assembler to put the pieces together, by another inspector, and so on. Each member of this hypothetical chain takes from the drawings information that he needs for his particular function. It is, therefore, easy to see why the drawings must be accurate and clear, free from ambiguities or misleading representations. Just as the years of written communication have led to rules and conventions, so years of manufacturing and production experience have led to drafting rules and conventions that help prevent errors.

4-2 ORTHOGRAPHIC PROJECTIONS

One of the most useful systems used by drafters to help assure accurate communication is orthographic projection.

Orthographic projections are views of an object taken at right angles to the object and arranged in specific relative positions on the drawing.

There is an infinite number of possible orthographic projections—there is an infinite number of ways to look at an object—but the views most commonly used are front, top, bottom, right side, left side, and rear (see Figure 4-1).

Six views are not generally required, for most objects may be completely defined in three views: front, top, and right (corresponding to height, width, and depth). Drafting convention calls for these views to be specifically placed on a drawing, and any variance is an error. Figure 4-2 shows the three views in correct position. Figure 4-3 shows two examples of positioning errors.

Each orthographic view is taken at right angles to the object it is defining. It is not a picture, such as an artist would draw, but a two-

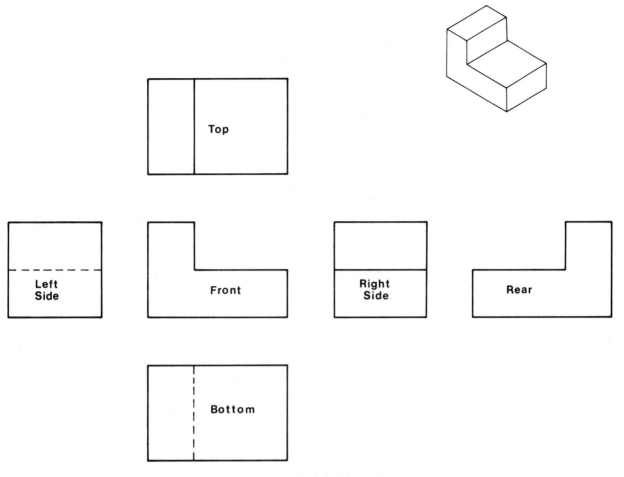

FIGURE 4-1 Front, top, bottom, right side, left side, and
rear views of an object.

FIGURE 4-2 Three views of an object located in correct
relative positions.

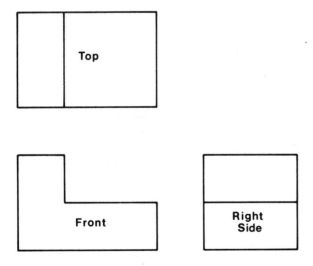

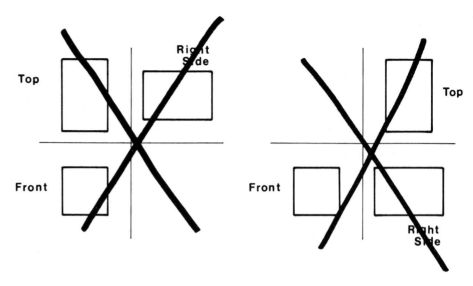

FIGURE 4-3 Orthographic views that are not positioned correctly.

dimensional representation which, for the sake of technical accuracy, has given up perspective. There is no shading and no attempt to "picture" the object. Each view presents only one face or one piece of the total information. Orthographic views are dependent on each other for a complete definition of the object. (There are objects that require fewer than three views, but these will be covered later.) In the top view in Figure 4-4, which surface is higher? There is no way to tell from this one orthographic view. Other views are needed before an answer may be given.

FIGURE 4-4 Given a top view of an object, which surface is higher? Other views are needed before an answer can be formulated.

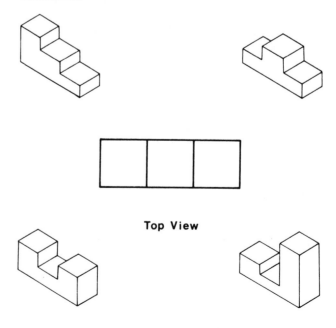

Top View

4-3   PRINCIPAL PLANE LINE

Drawings are divided into zones. Each zone contains one orthographic view along with all information pertinent to that view. The zones are separated by crossed (at 90°) construction lines called *principal plane lines* which are similar to a mathematical coordinate system. They are omitted on most finished drawings, but their presence is tacit. They will be included for the first problems in order to help establish the importance of the separation and relative position of views.

Principal plane lines are defined in Figure 4-5(a). Figure 4-5(b) shows how principal planes lines were initially developed.

FIGURE 4-5   (a) Principal plane lines; (b) how principal plane lines were initially developed.

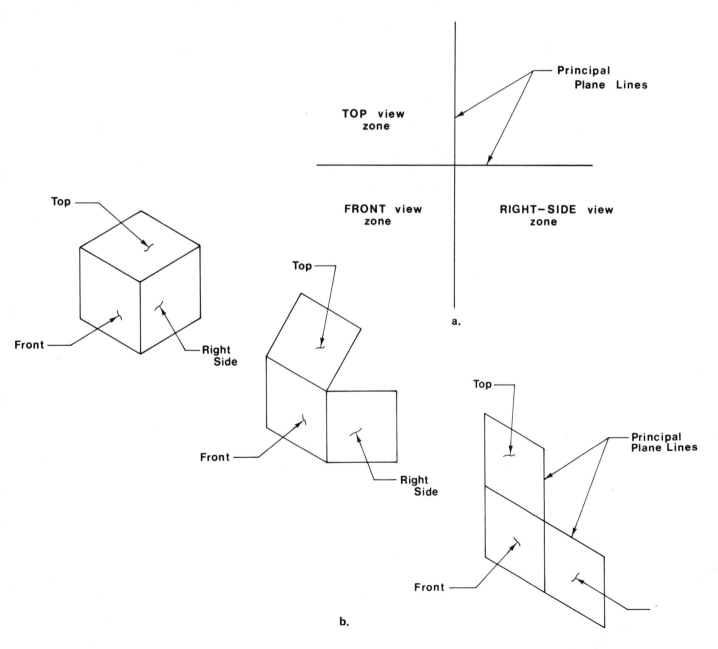

4-4 POINTS

Projection theory is the study of how to transfer information from one orthographic view to another. Often, two views of an object may be visualized, or parts of each view may be drawn, but the completed drawing remains clouded. Projection theory enables the bits and pieces to be used together to arrive at a finished drawing.

Reduced to its simplest form, projection theory may be used to transfer a single point from one view to another. Figure 4-6 presents the problem of finding the right side view of a point where the front and top views are given. Figure 4-7 shows the solution.

GIVEN: Front and top views of point 1.
PROBLEM: Draw the side view of point 1.

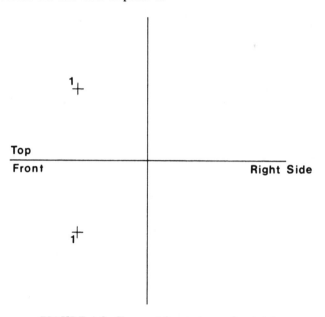

FIGURE 4-6   Top and front views of point 1.

SOLUTION:

FIGURE 4-7

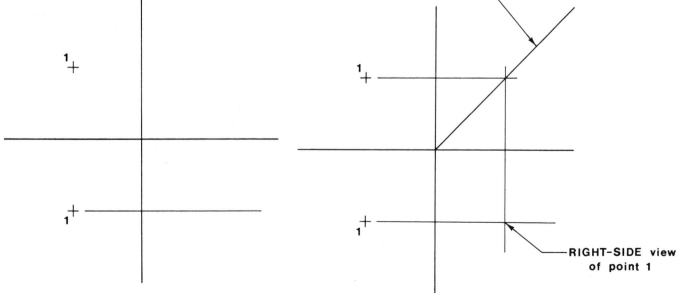

1. Project the front view of point 1 into the right-side view zone. This is done by drawing a horizontal construction line parallel to the horizontal principal plane line. The tendency here is to draw the projection line too short meaning extension may be required later. All we know at this time is that the right-side view is *somewhere* along the projection line.

2. Draw a line 45° up and to the right from the intersecting point of the principal plane lines. This is called a *mitre line*.

3. Project the top view of point 1 into the right-side view zone. This is done by drawing a horizontal construction line to the right, parallel to the horizontal principal plane line until it touches the 45° mitre line. When the projection line touches the mitre line, it turns the corner (that is, goes from horizontal to vertical). To continue the projection line, draw a vertical construction line, parallel to the vertical principal plane line, extending down into the right-side view zone. As in step 1, do not be stingy with the lead; draw the projection line through and beyond the horizontal projection line.

4. The intersection of the two projection lines is the right-side view of point 1. Label it.

Several additional points should be made before leaving this problem. The location of the front view of point 1 in relation to the top view is not random. The vertical line between the front and top views is parallel to the vertical principal plane line. Figure 4-8 shows three views of point 1 and the projection lines used to go from view to view. The point views and lines form a perfect rectangle (a four-sided figure with four right angles). This *projection rectangle* enables the drafter to find any third view of a point when given the two other views. This means that if we consider only three principal views (top, front, and right side), there are only three possible projection problems.

FIGURE 4-8   Projection rectangle.

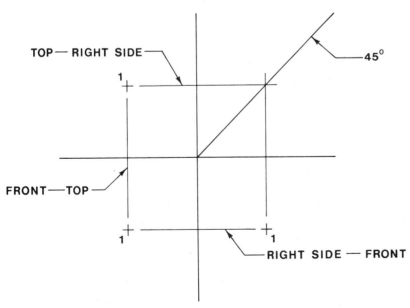

1. When you are given the front and top, draw the right side.
2. When you are given the front and right side, draw the top.
3. When you are given the top and right side, draw the front.

The sample problems in Figures 4-9 through 4-12 are examples of the other two possible point projection problems. Study them before proceeding to the next section.

GIVEN: Top and right-side views of point 2.
PROBLEM: Draw the front view.

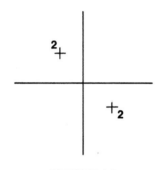

FIGURE 4-9

SOLUTION:

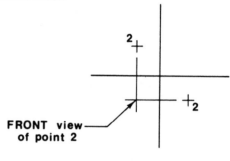

FIGURE 4-10

GIVEN: Front and right-side views of point 3.
PROBLEM: Draw the top view.

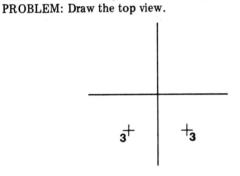

FIGURE 4-11

SOLUTION:

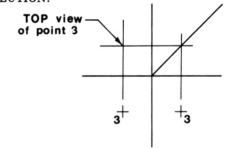

FIGURE 4-12

4-5 LINES

The projection of lines between views follows directly from point projection theory if we consider the axiom:

To a drafter, a line is a visible line that connects two
or more points.   Axiom 4-1

It follows then that lines may be projected by projecting the points that define them.

Figure 4-13 presents the problem of finding a right-side view when the front and top views are given. Figure 4-14 is the solution and was arrived at by the following:

1. Projecting point 1 into the right-side view (see Figure 4-1).
2. Projecting point 2 into the right-side view.
3. Connecting points 1 and 2 with an object line.

Step 3 is the right-side view of line 1-2.

GIVEN: Front and top views of line 1-2.
PROBLEM: Draw the side view.

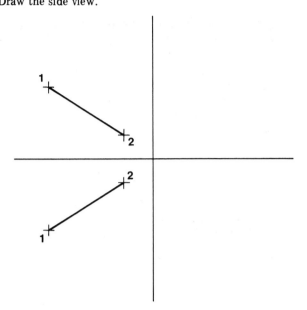

FIGURE 4-13

SOLUTION:

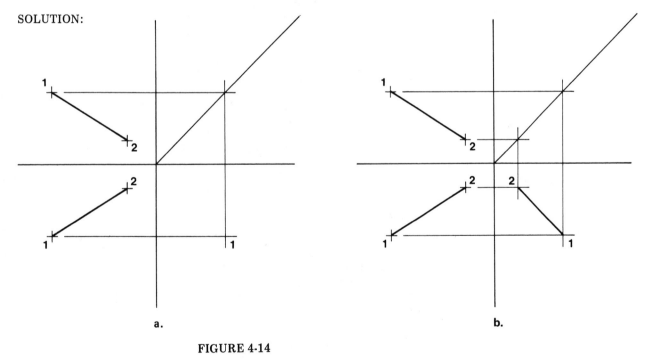

a.   b.

FIGURE 4-14

One aspect of line projection that could cause confusion is a double-point projection. This is clarified by the following axiom:

The end view of a straight line is a point (really a double point).   Axiom 4-2

Figure 4-15 is an example of a double-point projection. The solution (Figure 4-16) is derived exactly as shown in Figure 4-14, except for step 3. Points 1 and 2 cannot be joined by an object line because the line extends into the paper and therefore appears as a double point. This may be visualized if you hold a pencil horizontal to the ground and rotate it until you are looking directly at the point with the eraser end directly behind the point. If the point represents point 1 and if the eraser represents point 2, you now have a model of the end view of a line.

Figures 4-17 through 4-22 are samples of solved line projection problems. Study them before proceeding to the next section.

GIVEN: Front and top views of line 5-6.
PROBLEM: Draw the right-side view.

SOLUTION:

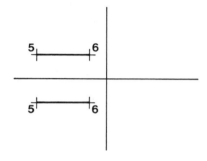

FIGURE 4-15   End view of a line.

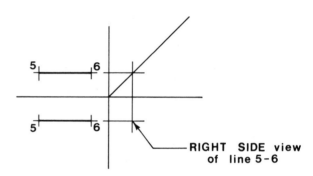

FIGURE 4-16

GIVEN: Front and right-side views of line 4-5.
PROBLEM: Draw the top view.

SOLUTION:

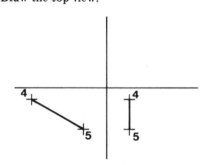

FIGURE 4-17

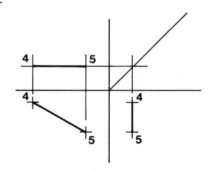

FIGURE 4-18

GIVEN: Front and right-side views of line 8-9.
PROBLEM: Draw the top view.

SOLUTION:

FIGURE 4-19

FIGURE 4-20

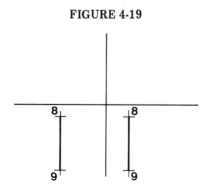

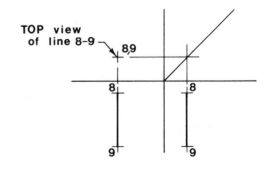

GIVEN: Top and right-side views of line 7-8.
PROBLEM: Draw the front view.

SOLUTION:

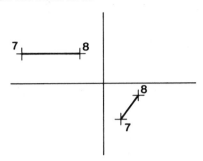

FIGURE 4-21

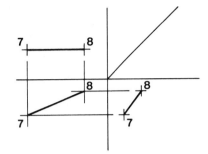

FIGURE 4-22

4-6   PLANES

As line projection theory was derived from point projection theory, so plane projection theory follows directly from line projection theory if we consider the following axiom:

To a drafter, a plane is the area enclosed within a
series of lines interconnected end to end.   Axiom 4-3

This differs from the geometric concept of planes in that it considers a plane a finite area, that is, an area with known boundaries.

Figure 4-23 gives the front and top views of plane 1-2-3-4 and asks for the right-side view. Figure 4-24 shows the solution, which was arrived at by the following:

1. Identify the lines that define the plane 1-2, 2-4, 4-3, and 3-1.
2. Project the individual points 1, 2, 3, and 4 into the right-side view (see Figure 3-7).
3. Draw in with object lines the lines that define the plane.

The lines drawn in step 3 define the right-side view of plane 1-2-3-4.

GIVEN: Front and top views of plane 1-2-3-4.
PROBLEM: Draw the right-side view.   SOLUTION:

FIGURE 4-23   FIGURE 4-24

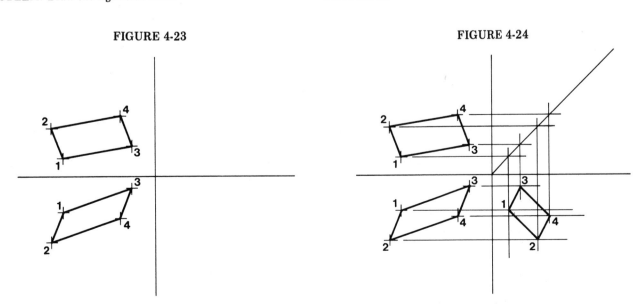

In line theory we found that the end view of a line was a double point. A similar situation appears in plane theory which is explained by the following axiom:

The end view of a plane is a line (really several lines
directly behind each other).   Axiom 4-4

This may be verified by holding a sheet of paper horizontal to the ground and rotating it until you are looking directly at one edge. Although it is a plane, the sheet appears as a line.

Figure 4-25 is a sample problem involving the end view of a plane. Points 1, 2, 3, and 4 are double points or end views of lines. Line 1-3 is located directly behind line 2-4 and is therefore hidden from view. Figure 4-26 is a good example of why orthographic views are dependent on each other to present a complete picture of an object. By itself, the right-side view is not only incomplete, it is also misleading.

GIVEN: Front and top views of plane 1-2-3-4.
PROBLEM: Draw the right-side view.

SOLUTION:

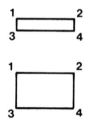

FIGURE 4-25

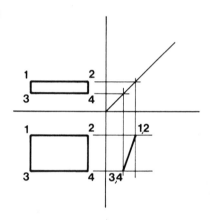

FIGURE 4-26

4-7 CURVES

So far we have considered only straight lines. Point, line, and plane projection theory may be extended to include curved lines if we consider the following axioms:

To a drafter, a curved line is a visible line connecting
three or more points which form a smooth, nonlinear
line.   Axiom 4-5

To a drafter, the accuracy of a curve is a function of
the number of points used to define the curve.   Axiom 4-6

To draw a perfectly accurate curve would require an infinite number of points. To do this is not only impossible, it is also impractical. Most curves may be very closely approximated by a finite number of points, and it is up to the drafter to determine which level of accuracy is required and how many points are needed to achieve this level.

Circles and perfect arcs are exceptions to the axioms because they may be drawn with perfect accuracy by using a compass.

Figures 4-27 and 4-28 are examples of curved-line projection problems and Figures 4-29 and 4-30 are examples of a plane with a curved edge. The solution to each of these problems is based on the concept of point project theory as shown in Figures 4-7 and 4-9 through 4-12.

GIVEN: Front and right-side views of curved line 1-2.
PROBLEM: Draw the top view.

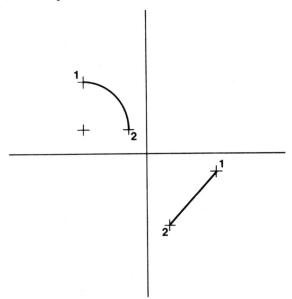

FIGURE 4-27

SOLUTION:

FIGURE 4-28

a.

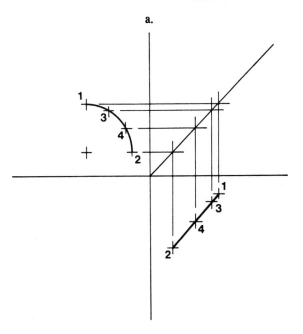

b.

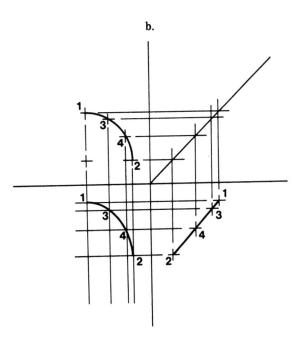

GIVEN: Front and top views of plane 1-2-3-4-5.
PROBLEM: Draw the side view.

SOLUTION:

FIGURE 4-29

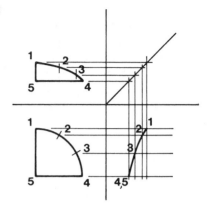

FIGURE 4-30

PROBLEMS

P4-1 Draw three views (front, top, and side) of the points, lines,
through and planes in each figure. Include principal plane lines, mitre
P4-5 lines, and projection lines for each problem. Each square on
the grid is 0.20 × 0.20.

FIGURE P4-1

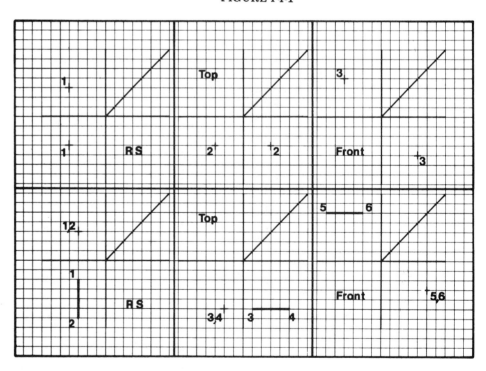

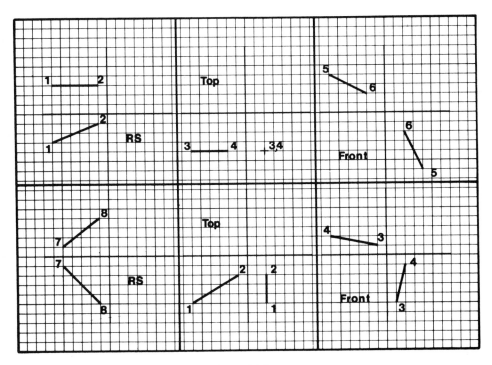

FIGURE P4-2

FIGURE P4-3

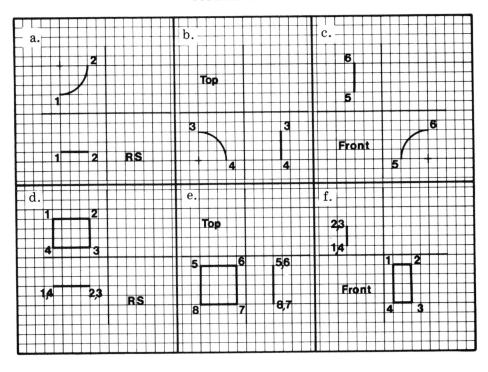

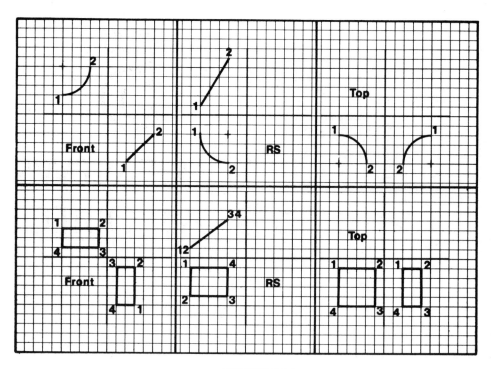

FIGURE P4-4

FIGURE P4-5

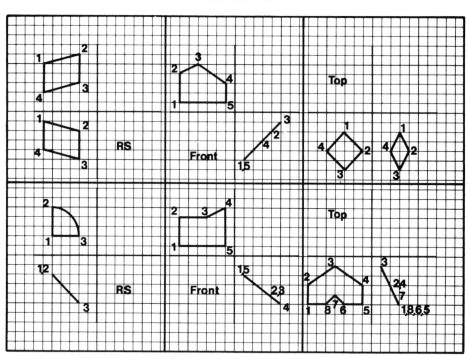

5

THREE VIEWS
OF AN OBJECT

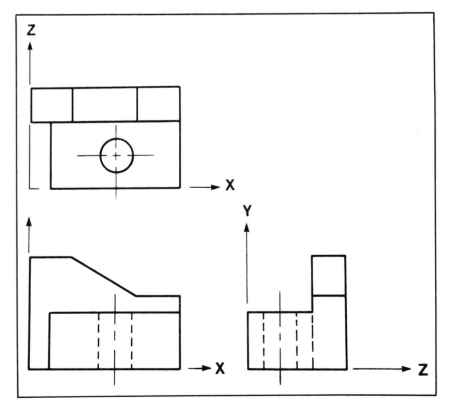

FIGURE 5-0

5-1 INTRODUCTION

In this chapter we extend the projection theory concepts of Chapter 4 to cover three-dimensional objects. The basic three views (front, top, and right side) and their relative locations on the drawing are the same for three-dimensional objects as they are for points, lines, and planes. Similarly, the techniques for projecting information from one view to another remain exactly the same. As we consider lines to be defined by points and we consider planes to be defined by lines, so we may consider three-dimensional objects to be defined by planes and, therefore, directly apply projection theory.

We also introduce in this chapter the concept of object visualization. *Object visualization* is the ability to mentally picture an object in three dimensions when only orthographic views are given and to mentally visualize the orthographic views of an object when only a three-dimensional picture is given. It is an important skill for a drafter to develop. Each sample problem in this and the next six chapters will include both the orthographic views and a three-dimensional drawing, called an *isometric drawing*, of the objects to be studied so that an understanding of the object visualization may be developed.

5-2 NORMAL SURFACES

Figure 5-1 shows an object and a three-view orthographic drawing of that object. All surfaces in the object are normal, that is, at 90° to each other. The principal plane lines and the projection lines have been included, and points 1, 2, 3, 4, 5, and 6 have been defined.

Planes 1-2-3-4 and 2-3-5-6 have been numbered to demonstrate the application of projection theory to objects. Projection theory is directly applicable to three-dimensional objects if we consider the following axiom:

To a drafter, an object is a volume enclosed within a
series of interconnected planes.

Axiom 5-1

As we are able to analyze lines from points and analyze planes from lines, so we are able to analyze objects from planes. Planes 1-2-3-4 and 2-3-5-6 are analyzed separately in Figure 5-2. All other surfaces that make up the object may be analyzed in the same way and then combined into a composite of planes which in turn form the three views of the object.

Normally, drafters do not number all points on an object because they see their finished drawings mentally before drawing them. This is not always true, however, because not every object can be solved men-

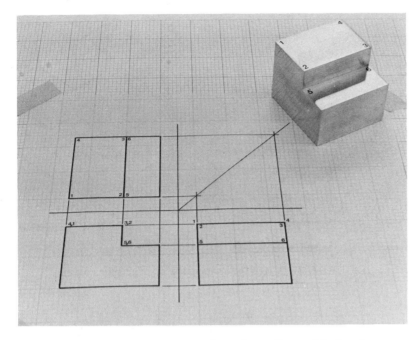

FIGURE 5-1 Object and a three-view orthographic drawing of that object.

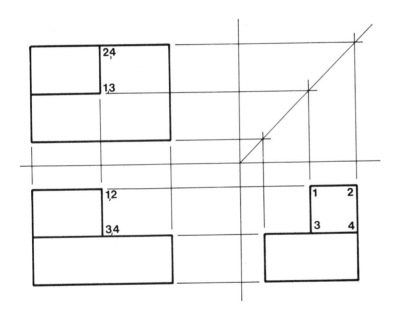

FIGURE 5-2 Object and a three-view orthographic drawing of that object.

tally. Thus drafters often use projection theory to help them derive and check surfaces about which they are unsure. Let us assume, for example, that surface 1-2-3-4 in Figure 5-2 has caused confusion and that we have numbered what we feel are the correct three views. We now wish to check our work.

To check the proposed solution, start with point 1 and draw in the projection rectangle verifying the indicated locations of point 1. Do

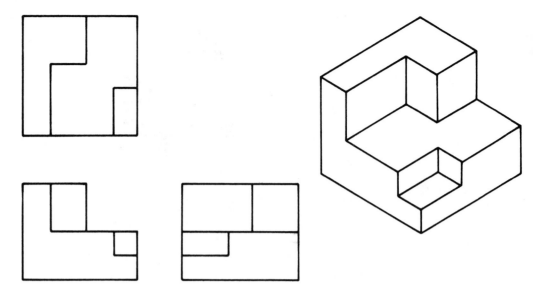

FIGURE 5-3 Object and a three-view orthographic drawing of that object.

the same with line 1-2 and then with surface 1-2-3-4. All points, lines, and the plane check. Therefore, the drawn solution is correct.

Figure 5-3 is another example of an object containing all normal surfaces.

5-3 HIDDEN LINES

Most objects contain lines (edges) that cannot be seen in all three views. The slot in Figure 5-4 appears directly in the top and right-side views, but it is hidden in the front view. We must somehow represent the slot in the front view to ensure that all views are consistent in the information they present. We do this by using hidden lines.

FIGURE 5-4 Example of an object whose front view contains a hidden line.

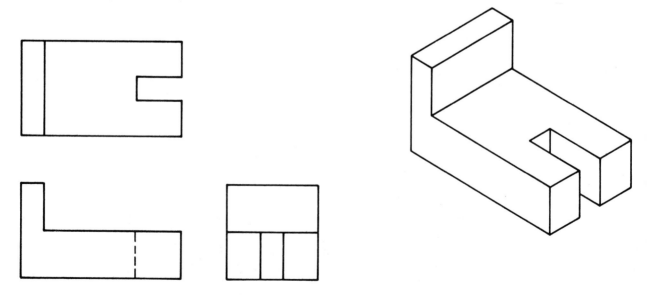

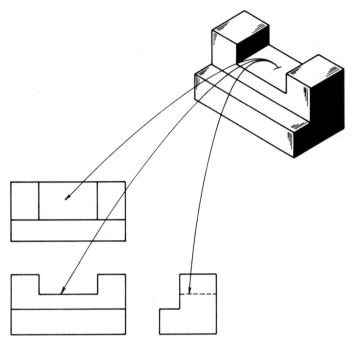

FIGURE 5-5 Pattern for hidden lines.

FIGURE 5-6 Hidden-line configuration.

Hidden lines are lines used to represent edges of an object that cannot be directly seen.　　　　　　　　　　　　Axiom 5-2

The hidden lines in the front view of the object shown in Figure 5-4 represent the horizontal surface of the slot. Figure 5-5 shows an object that contains a hidden line in its side view.

Hidden lines are drawn by using dashes as explained in Figure 5-6. The actual length of the dashes may vary according to the situation as long as a 4-to-1 ratio is maintained between the dashes and the intermittent spaces. Since hidden lines are not as dark or as thick as object lines, you should be careful to make sure that there is a noticeable difference between object lines and hidden lines. See Chapter 2 for further definition of kinds of lines.

There are three rules that must be followed when drawing hidden lines. They have been developed to prevent confusion and misunderstanding in the use and interpretation of hidden lines. Figure 5-7 illustrates the rules.

Rule 1: Do not continue an object line into a hidden line. Always allow a small (1/16) gap.　　　　　　　　Axiom 5-3

FIGURE 5-7(a) Do not continue a visible line into a hidden line; leave a gap.

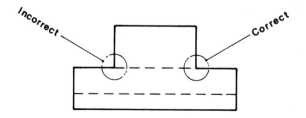

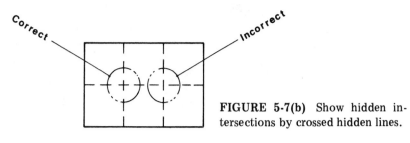

FIGURE 5-7(b) Show hidden intersections by crossed hidden lines.

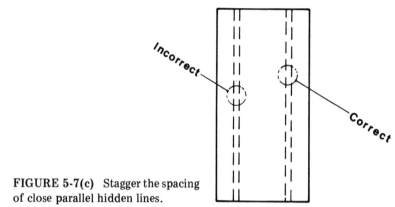

FIGURE 5-7(c) Stagger the spacing of close parallel hidden lines.

Rule 2: Show hidden corners as an intersection of hidden lines, thereby specifically defining the location of the corner.   Axiom 5-4

Rule 3: Never draw parallel hidden lines with equal-length dashes and spaces. Stagger the lengths so that each line is distinctive.   Axiom 5-5

Figures 5-8 through 5-11 are further examples of hidden-line problems.

FIGURE 5-8 Example of an object whose orthographic views contain hidden lines.

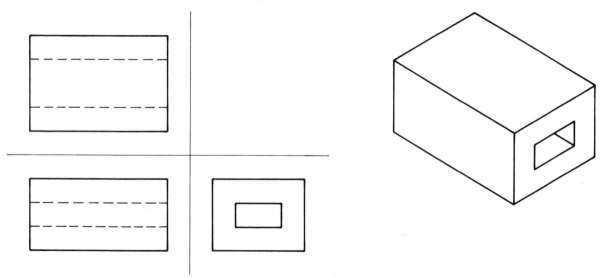

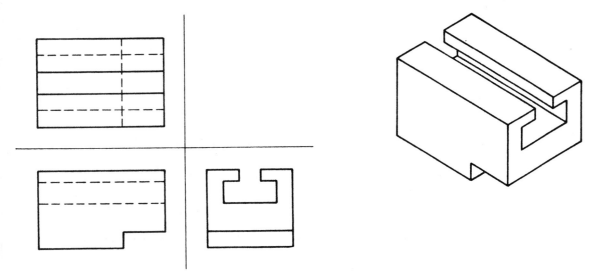

FIGURE 5-9 Example of an object whose orthographic views contain hidden lines.

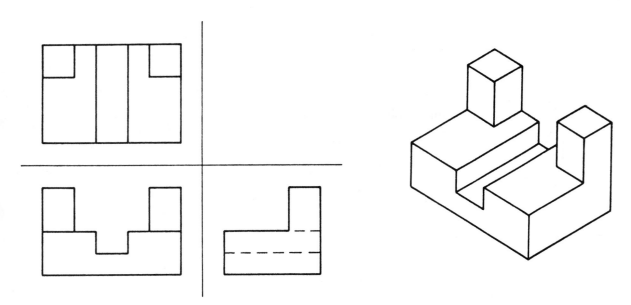

FIGURE 5-10 Example of an object whose orthographic views contain hidden lines.

FIGURE 5-11 Example of an object whose orthographic views contain hidden lines.

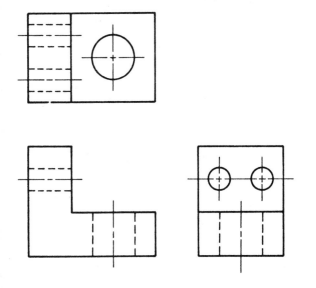

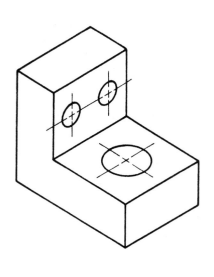

5-4 INCLINED SURFACES

Figure 5-12 shows an object that has an inclined surface 1-2-3-4. An *inclined surface* is one that is parallel to one, but not both, principal plane lines. Note that the top and right-side views are approximately the same as those shown for the example in Figure 5-1 and note the incline of plane 1-2-3-4 may only be seen in the front view. This kind of visual ambiguity is unavoidable in orthographic views, and as shown here it emphasizes the importance of using all orthographic views together to form a final solution to the problem.

Figures 5-13, 5-14, and 5-15 are other sample problems that include inclined surfaces.

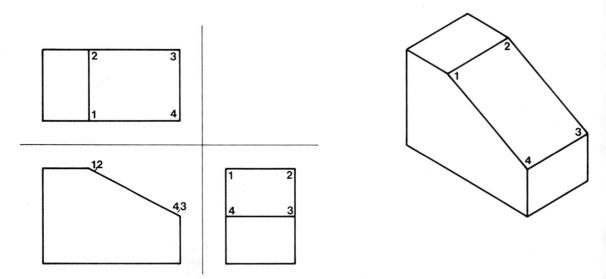

FIGURE 5-12   Object with an inclined surface 1-2-3-4.

FIGURE 5-13   Object with an inclined surface.

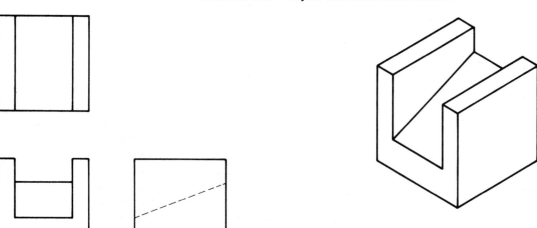

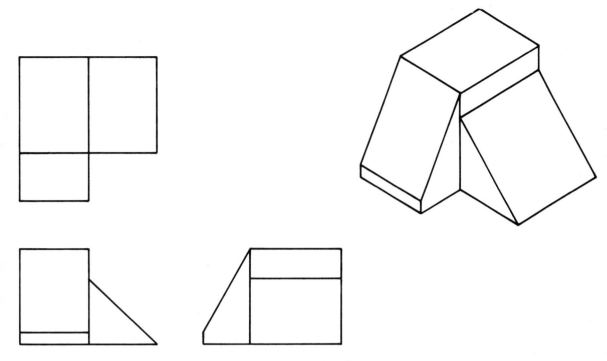

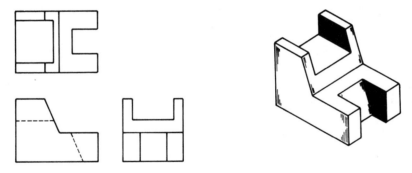

FIGURE 5-14   Object with inclined surfaces.

FIGURE 5-15   Object with inclined surfaces.

5-5   CURVED SURFACES

Figure 5-16 shows an object that has a curved surface. A *curved surface* is one that appears as a part of a circle (an arc of constant radius) in one of the orthographic views. Curved surfaces are similar to slanted surfaces in that they tend to generate ambiguous orthographic views.

A unique characteristic of curved surfaces is the tangency line. Surface 1-2-3-4 in Figure 5-16 contains a tangency line 5-6 represented by a phantom line. A tangency line represents the location at which the round portion of surface 1-2-3-4 flairs into (becomes tangent to) the flat horizontal portion. Because there is no edge here, a line would not be drawn in any of the views. Figure 5-17 shows an object in which the curved surface does form an edge with the lines labeled 7-8 and 9-10; thus it requires an object line. Without exception, you may always draw a visible line when the round surface forms an edge with the other surfaces.

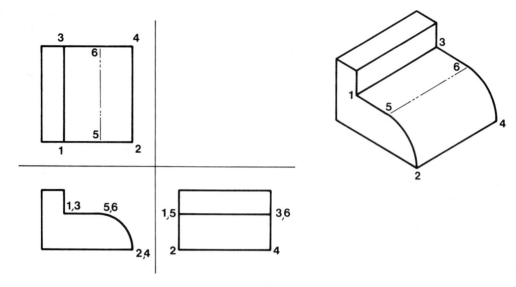

FIGURE 5-16 Object with a curved surface.

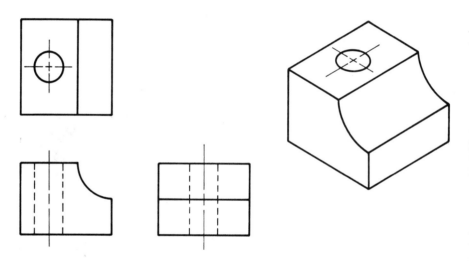

FIGURE 5-17 Object with a curved surface.

You cannot always omit an object line if a physical edge does not exist. Figure 5-18 shows six examples of objects that contain rounded surfaces. As a rule, if a curved surface changes direction (goes from concave to convex, or vice versa) such that any part of the curved surface becomes parallel to one of the principal plane lines, a line is required. Note in Figure 5-18 that the top views of objects B and C contain object lines that represent a directional change of the curved surface. In object E, two lines are generated; one is solid and the other is hidden. In object D, a hidden line is required in the side view.

Figure 5-19 shows another example of orthographic views that contain rounded surfaces.

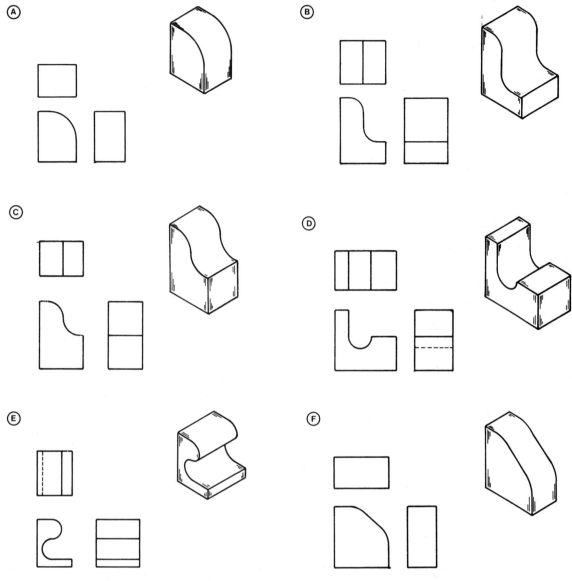

FIGURE 5-18   Six objects with curved surfaces.

FIGURE 5-19   Object with curved surfaces.

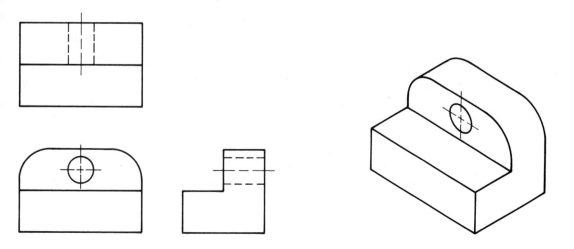

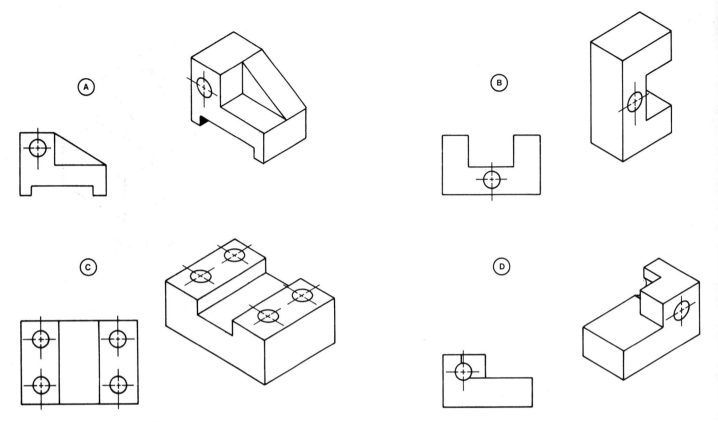

FIGURE 5-20 Front views that visually most completely describe the object.

5-6 CHOOSING A FRONT VIEW

When preparing orthographic views, it is important to choose a front view that visually most completely describes the object. Figure 5-20 shows four objects and appropriate front views. In each example, the front view was chosen because it most clearly describes the most individual features. No single view will describe all features of an object, but the front view should show as many features as possible.

5-7 SKETCHING

Before drafters begin the actual drawing of a new assignment, they usually make a sketch of the object involved. They then study the sketch and try to identify any problems that could arise when making the drawing. If any problems are found, they rework the sketch until the problems are solved and the sketch has become a clear, well-understood picture of the future drawing. Drafters take the time to create good sketches because it is much easier to correct freehand sketches than to correct finished drawings. The time spent sketching is more than regained when creating a drawing because they avoid the problems they found and corrected while making the sketch.

Of course, learning to sketch is easier for those who have artistic ability, but anyone can learn to sketch. The following hints are offered to make it easier for you to learn to sketch.

Making Sketches of Orthographic Views (see Figures 5-21 and 5-22)

GIVEN: An object.

PROBLEM: Sketch the front, top, and right-side views.

SOLUTION:

1. Use grid paper, graph paper, quadrapads, and so on. This kind of paper will help you to establish an approximate scale and thereby keep your sketches fairly proportioned. It will also help you to keep your lines straight.

2. Lightly sketch the overall shape of the object as would be seen in the three basic orthographic views.

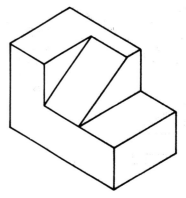

FIGURE 5-21

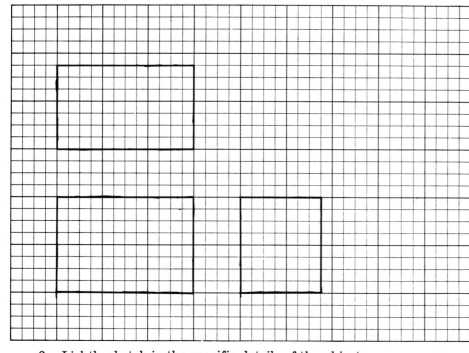

FIGURE 5-22(a)

3. Lightly sketch in the specific details of the object.

FIGURE 5-22(b)

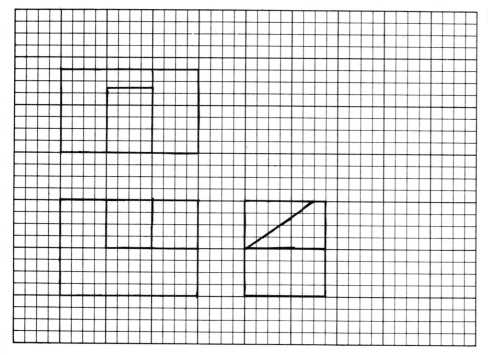

4. When the desired shape is completed, darken in the important lines by using heavy, bold strokes.

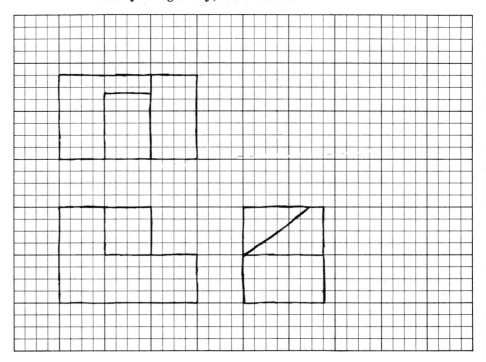

FIGURE 5-22(c)

Making a Picture (Isometric) Sketch (see Figures 5-23 and 5-24)

GIVEN: Three views of an object.
PROBLEM: Sketch an isometric picture of the object.

FIGURE 5-23

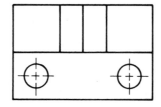

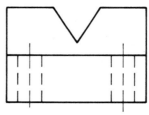
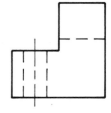

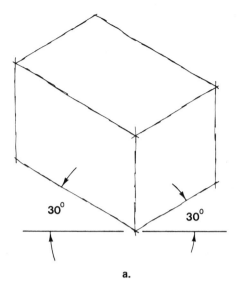

a.

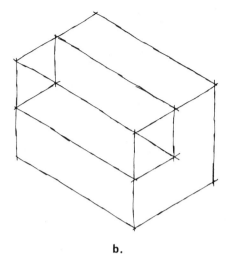

b.

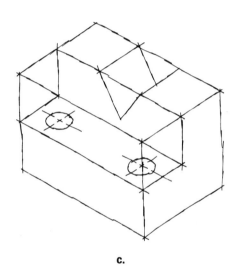

c.

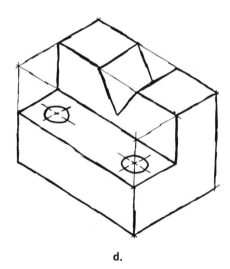

d.

FIGURE 5-24

SOLUTION:

1. Draw a block whose length, width, and height are of approximately the same proportions as those of the object to be sketched. Make the receding lines of the box 30°.

2. Lightly sketch in the specific details of the object.

3. When the desired shape is completed, darken in the important lines by using heavy, bold strokes.

Figure 5-25 shows an example of freehand sketches of the orthographic views for an object. No grid background was used. Instead, rectangular outlines were initially drawn for the front, top, and side views using the proportions of the object.

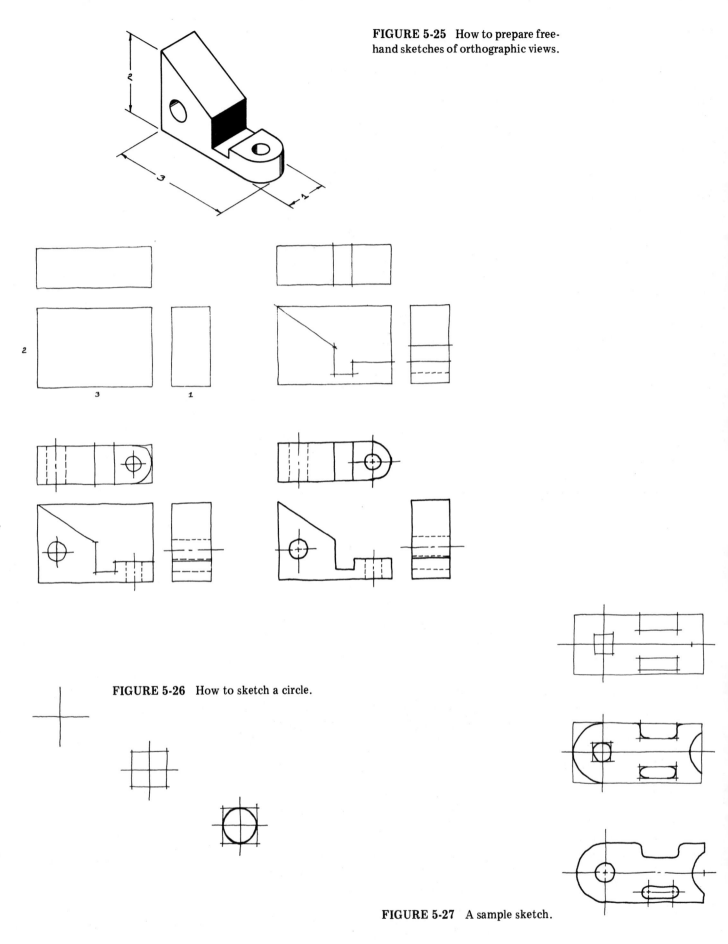

FIGURE 5-25 How to prepare free-hand sketches of orthographic views.

FIGURE 5-26 How to sketch a circle.

FIGURE 5-27 A sample sketch.

Round shapes are sometimes difficult to sketch. Figure 5-26 shows a procedure that should be helpful when sketching round shapes. Start with the centerlines of the shape, then outline the shape as if it were square, that is, made entirely of straight lines. The round shape can then be sketched using the intersections of the centerlines and rectangular outline as reference points. Figure 5-27 is a further example.

5-8 VISUALIZATION TECHNIQUES

Visualizing an object in three dimensions, given only the orthographic views, has always been a problem for drafters. Drawing and sketching experience and good depth perception help, but there are always those problems that just "cannot be seen." Two techniques used by drafters to help visualize difficult problems are model building and surface coloring.

Models offer the best visualization aids because they themselves are three-dimensional objects, but models are usually expensive and time consuming to build. Figure 5-28 shows examples of some well-constructed models. To overcome the expense and time constraints, some drafters make models out of children's modeling clay. Figure 5-28 shows an example. Clay models are not meant to be exact-scaled duplications, but rather approximate representations made to help a drafter visualize the object being drawn; thus the quality of clay models may vary according to personal requirements and situations.

FIGURE 5-28   Models used to help visualize. The model in the center is made from children's modeling clay.

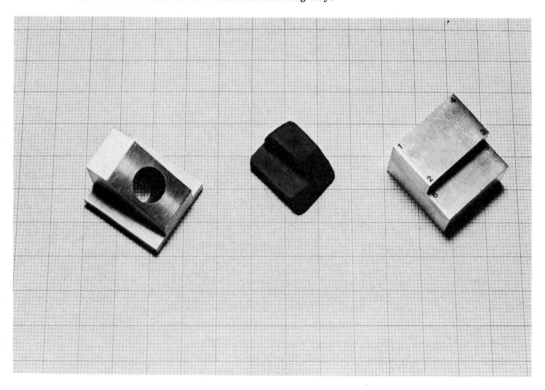

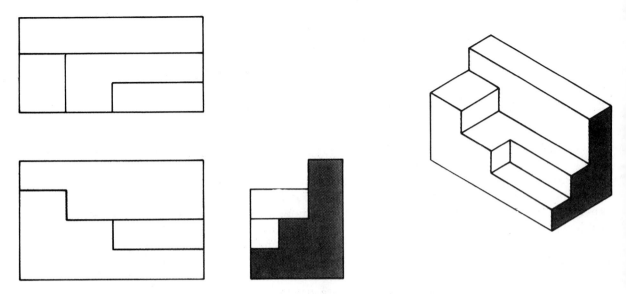

FIGURE 5-29 Example of surface shading.

Figure 5-29 shows an example of surface coloring. Drafters generally color by using different colors (red, blue, and so on), but the example in Figure 5-29 was done by using various shades of gray. By coloring a surface with the same color in all the different views, the surface may be more easily identified in the various views and therefore more easily visualized.

PROBLEMS

P5-1 through P5-22 Draw or sketch, as assigned by your instructor, three views (front, top, and right side) of each object. Each triangle in the grid pattern is 0.20 on a side, except for Problems 5-1 and 5-3, which use a 0.25-on-a-side grid pattern.

FIGURE P5-1

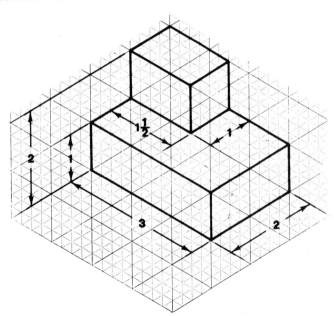

FIGURE P5-2

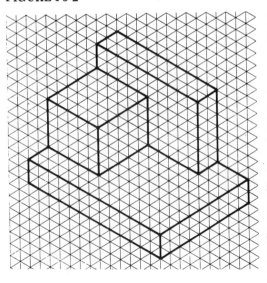

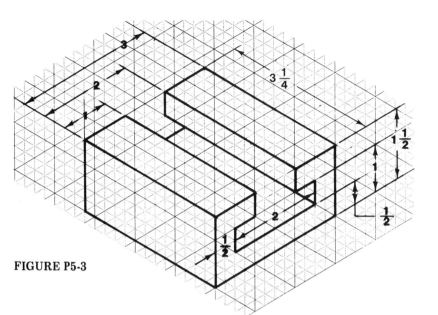

FIGURE P5-3

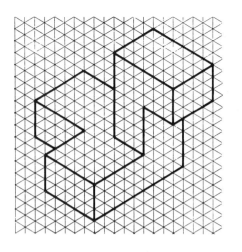

FIGURE P5-4

FIGURE P5-5

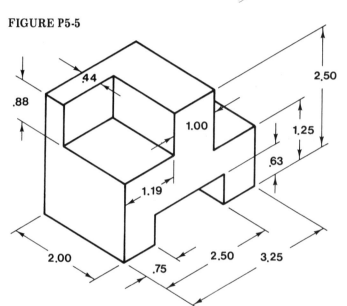

FIGURE P5-6

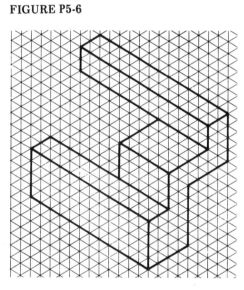

FIGURE P5-7

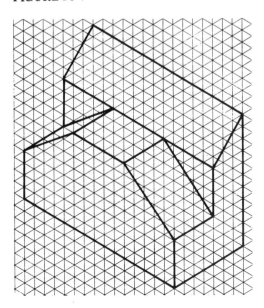

FIGURE P5-8

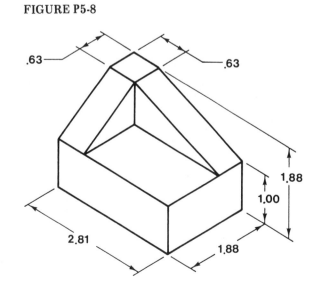

FIGURE P5-9

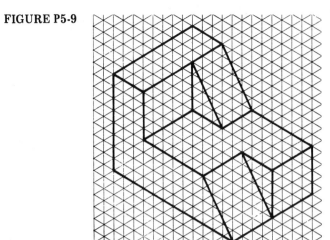

FIGURE P5-10

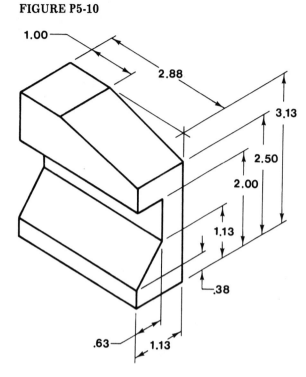

FIGURE P5-11

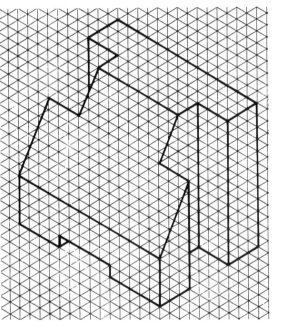

FIGURE P5-12

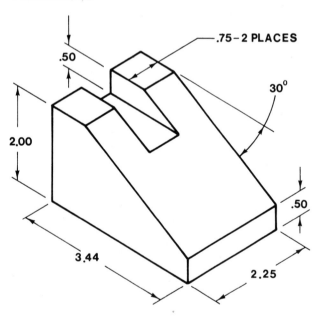

FIGURE P5-13

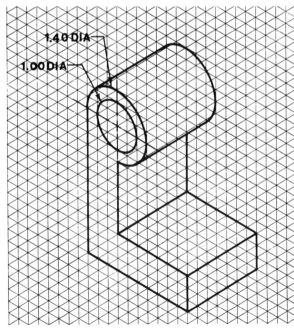

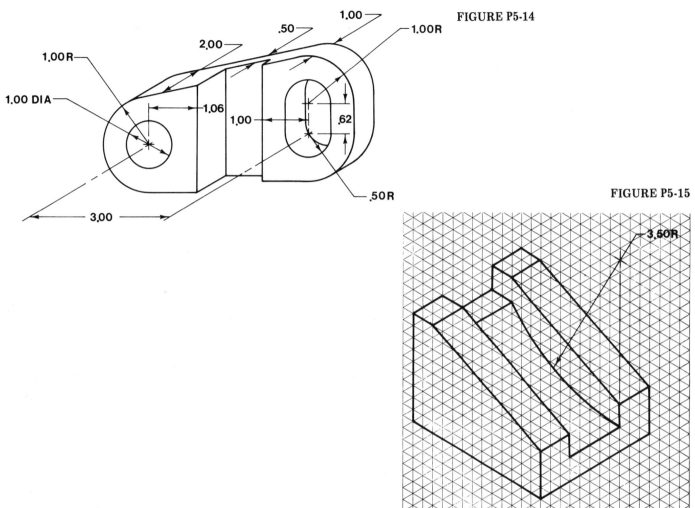

FIGURE P5-14

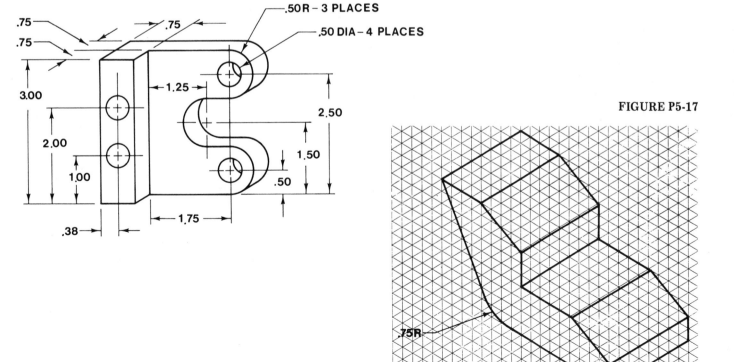

FIGURE P5-15

FIGURE P5-16

FIGURE P5-17

FIGURE P5-18

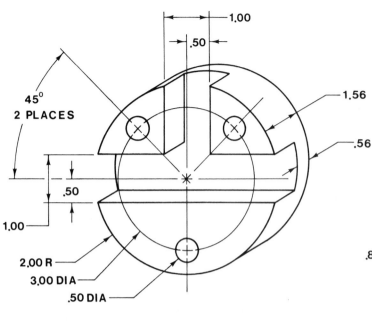

FIGURE P5-19

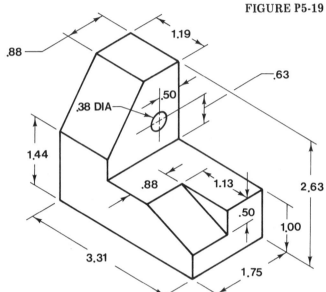

FIGURE P5-20

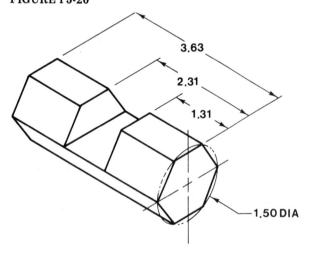

FIGURE P5-21

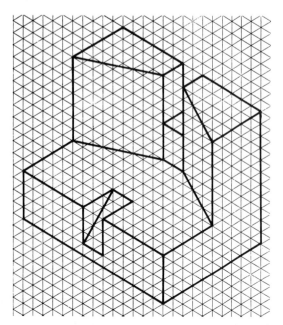

FIGURE P5-22

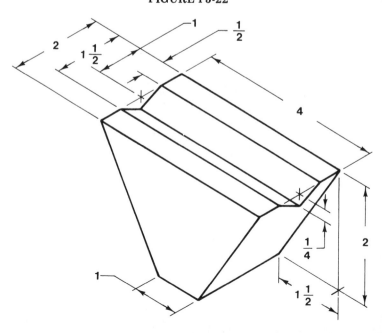

P5-23 For each figure, redraw the two views given and add the third
through as required. If assigned, prepare a freehand three-dimensional
P5-28 sketch of the object. Each square on the grid pattern is 0.20
per side.

Top

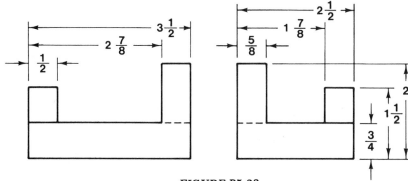

FIGURE P5-23

FIGURE P5-24

FIGURE P5-25

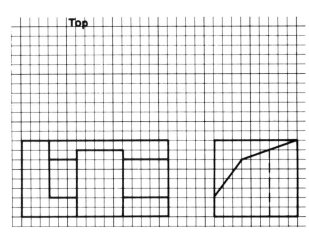

Top

FIGURE P5-26

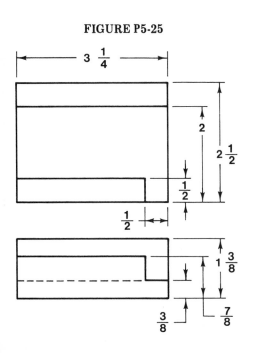

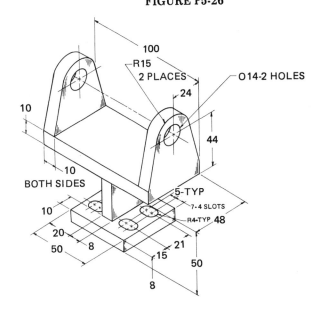

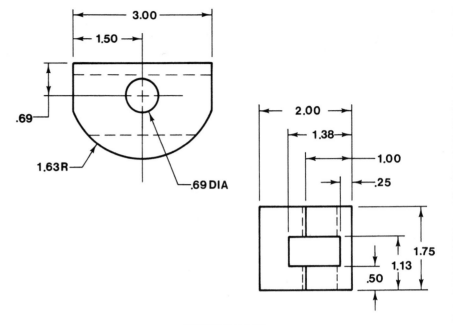

FIGURE P5-27

FIGURE P5-28   All dimensions are in millimeters.

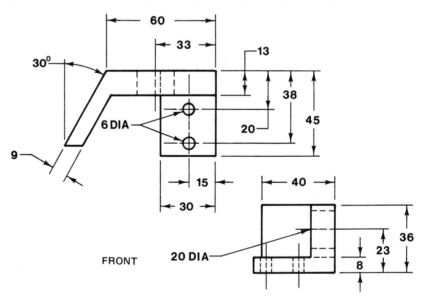

P5-29 Prepare a front, top, and right side view of the following
through objects.
P5-40

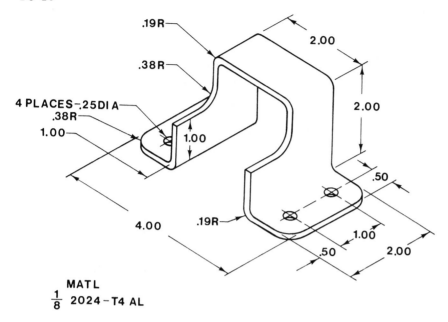

MATL
$\frac{1}{8}$ 2024-T4 AL

FIGURE P5-29

FIGURE P5-30

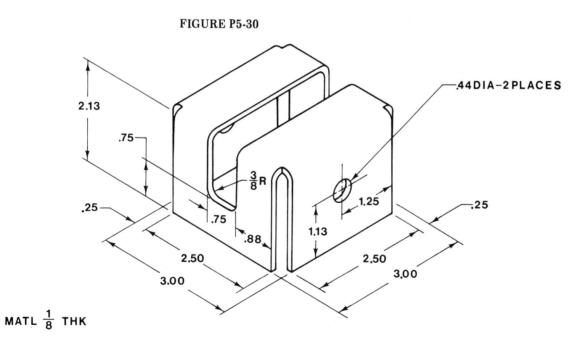

MATL $\frac{1}{8}$ THK

ALL INSIDE BEND RADII $\frac{3}{16}$

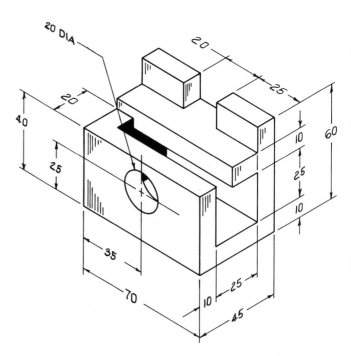

FIGURE P5-31   All dimensions are in millimeters.

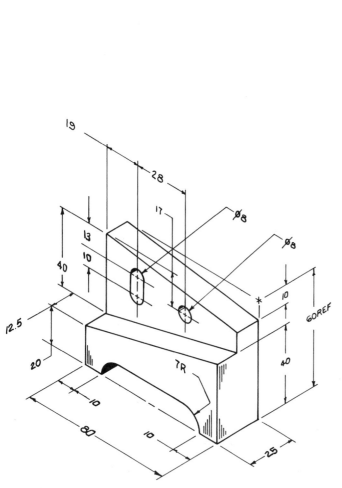

FIGURE P5-32   All dimensions are in millimeters.

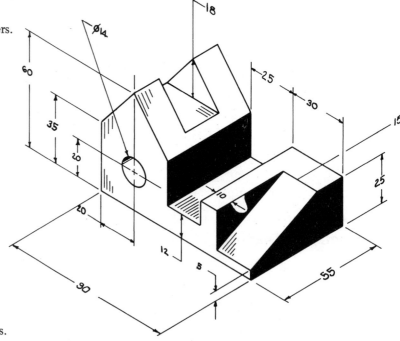

FIGURE P5-33   All dimensions are in millimeters.

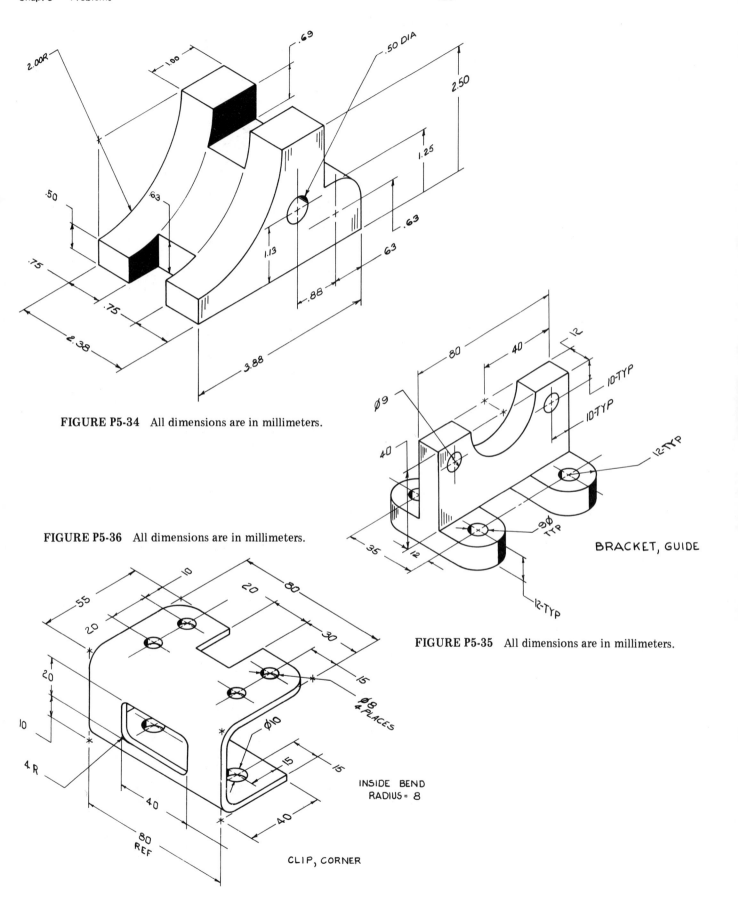

FIGURE P5-34   All dimensions are in millimeters.

FIGURE P5-36   All dimensions are in millimeters.

BRACKET, GUIDE

FIGURE P5-35   All dimensions are in millimeters.

INSIDE BEND
RADIUS = 8

CLIP, CORNER

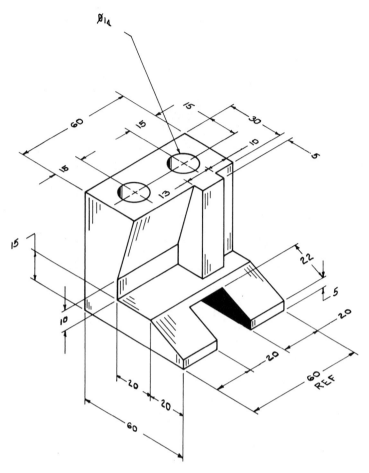

FIGURE P5-37   All dimensions are in millimeters.

FIGURE P5-38   All dimensions are in millimeters.

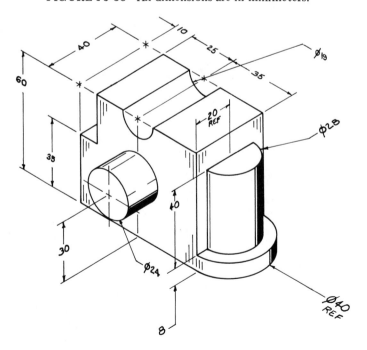

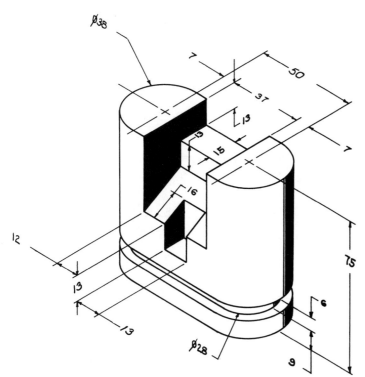

FIGURE P5-39 All dimensions are in millimeters.

FIGURE P5-40 All dimensions are in millimeters.

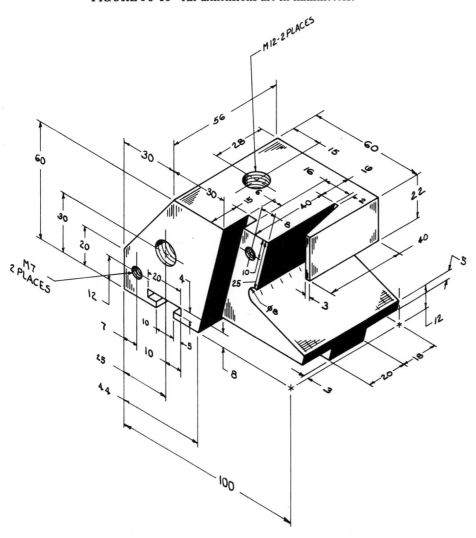

6

DIMENSIONS
AND TOLERANCES

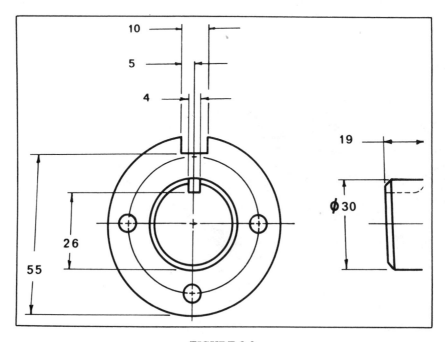

FIGURE 6-0

6-1 INTRODUCTION

This chapter explains and illustrates dimensioning and tolerancing. The picture portion of a drawing defines the shape of the object, the dimensions define the size, and the tolerances define the amount of variance permitted in the size. All three pieces of information are needed to form a clear, understandable, manufacturable drawing.

To help you gain an understanding of the relationships between size, shape, and tolerance, look at Figure 6-1. What is the height of the car? Is it full size or is it merely a model? We may get some approximation of the height by comparing the height of the car with the height of the woman. If we use the woman for our scale, we may say that the car is a little less than one-half a woman height. Here the picture gives us shape and the woman gives us an approximate size—but what about the tolerance? How tall is the woman? Is she wearing high-heeled shoes or is she standing on a box? For a more accurate answer, we need a more accurate scale.

The post beside the woman has been calibrated into 6 inch-intervals. Further, it has been cut off and labeled 35-1/2 inches. If the post is our

FIGURE 6-1   Courtesy of General Motors.

scale and the 35-1/2 inches-label is our dimension, we are assured of a more accurate measurement. As before, since the accuracy of our final measurement depends on the accuracy of our scale, we would pick a scale that satisfies our specific tolerance requirement. If we just want to know about how high the car is, the girl would be sufficient. If we want to know within an inch, the post dimension would probably be acceptable. If we want a more accurate answer, we would have to use a more accurate scale.

As you read through this chapter, remember that dimensions are the most important part of any drawing. Always try to dimension your drawings clearly, concisely, and in an easily understandable manner.

6-2 EXTENSION LINES, DIMENSION LINES, LEADER LINES, AND ARROWHEADS

Dimensions are placed on a drawing by using a system of extension lines, dimension lines, leader lines, and arrowheads. Figure 6-2 illustrates how these various kinds of lines are used for dimensioning. The lines are defined as follows:

EXTENSION LINES: used to indicate the extension of an edge or point to a location outside the part outline.*
DIMENSION LINES: show the direction and extent of a dimension.*
LEADER LINES: used to direct an expression, in note form, to the intended place on the drawing. The leader line should terminate in an arrowhead or dot.*
ARROWHEADS: used to indicate the ends of dimension lines and the ends of some leader lines.* Arrowheads are drawn as shown in Figure 6-3.

FIGURE 6-2   Extension, dimension, and leader lines.

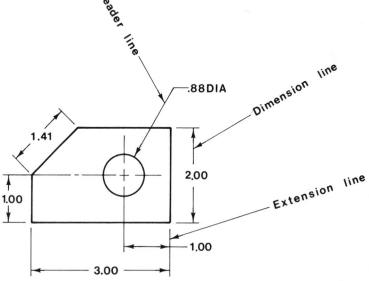

*Extracted from American Drafting Standards, Line Conventions, Sectioning, and Lettering (ANSI Y14.5-1982) with the permission of the publisher, The American Society of Mechanical Engineers, United Engineering Center, 345 East 47th Street, New York, NY 10017.

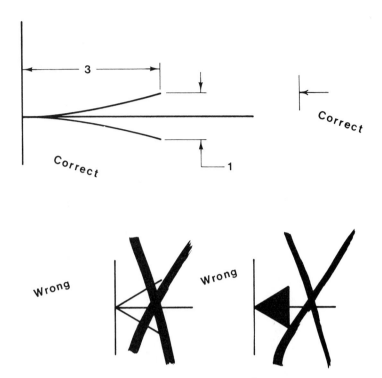

FIGURE 6-3   Arrowheads.

6-3   LOCATING AND PRESENTING DIMENSIONS

How do you locate and present dimensions on a drawing so that they may be easily and unmistakably understood? Unfortunately, there is no one answer to this question. Each drawing must be dimensioned according to its individual requirements, and what is acceptable in one situation may not be acceptable in another situation. Learning how to locate and how to present dimensions depends a great deal on drawing experience, but there are some general guidelines that may be followed. These guidelines are presented below and are illustrated in Figure 6-4.

    (a) Dimension by using extension, dimension, and leader lines placed neatly around the various views of the object. Place dimensions so that your reader will have no difficulty understanding which surface or which edge you are defining.

FIGURE 6-4(a)

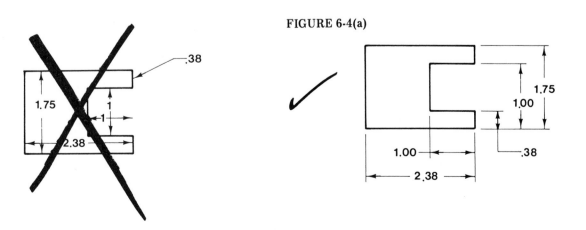

(b) Be sure that the size of the object is completely defined and that no surfaces or edges are left out.

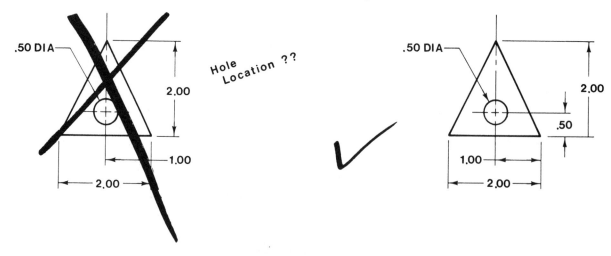

FIGURE 6-4(b)

(c) Always keep dimensions at one constant height. 1/8 or 3/16 is the generally accepted height, although a larger height may be used in some cases (title blocks, page numbers, and so on). Letters and numbers should never be less than 1/8 in height.

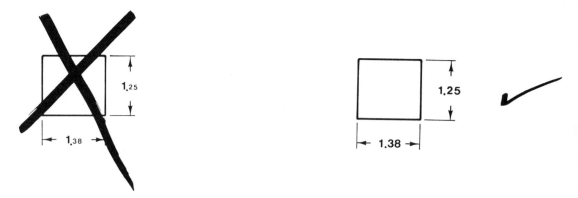

FIGURE 6-4(c)

(d) Do not squeeze dimensions into small spaces and angles. Undersized dimensions are difficult to read.

FIGURE 6-4(d)

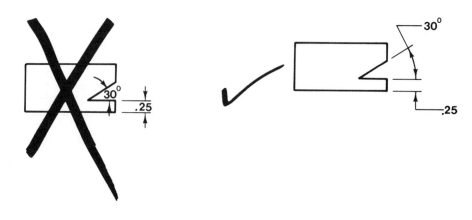

(e) Unless it is absolutely necessary, do not put any dimensions within the visible lines of the object being defined. You will never know when or how a drawing may have to be changed. It is important that you realize that drawing changes are not necessarily the result of errors. Customer requirements may change, designs may be modified, a new machine may be added to your company's manufacturing process, and so on. Any one of these reasons, and many more, could necessitate drawing changes.

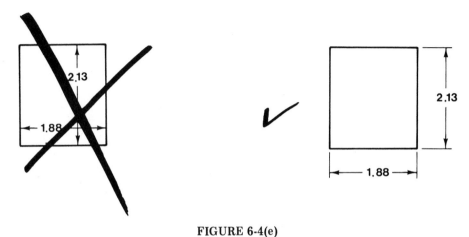

FIGURE 6-4(e)

(f) Do not overdimension. Too many dimensions are as confusing as too few dimensions. A common mistake is to double dimension, that is, to dimension the same distance twice on the same drawing. One dimension per distance is sufficient.

FIGURE 6-4(f)

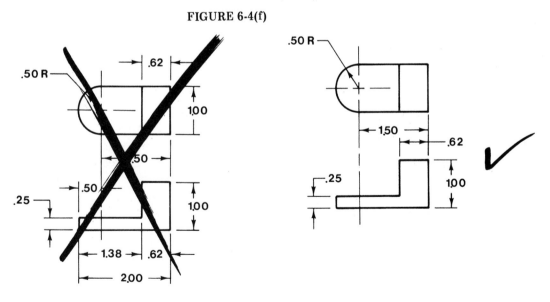

(g) Do not place dimensions too close to the object to be defined. A dimension line should never be closer than 3/8 to the object.

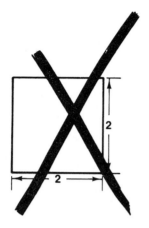

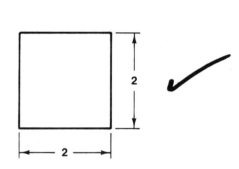

FIGURE 6-4(g)

(h) Leader lines should all be at the same angle. This will tend to give the drawing a more organized appearance.

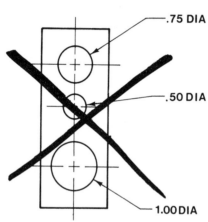

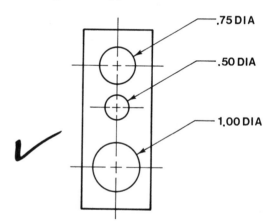

FIGURE 6-4(h)

(i) Space dimensions evenly. This not only gives the drawing a well-organized appearance, but it also makes the dimensions much easier to read.

FIGURE 6-4(i)

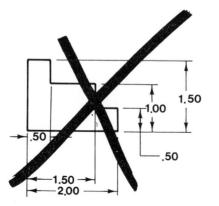

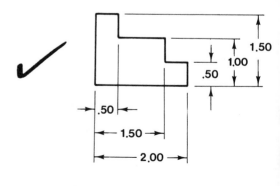

(j) Leader lines should not change directions until after they have extended beyond the outside edge of the object and beyond any dimension or extension lines. Leader lines should always end in a short horizontal section that will guide the reader's eye into the appropriate note.

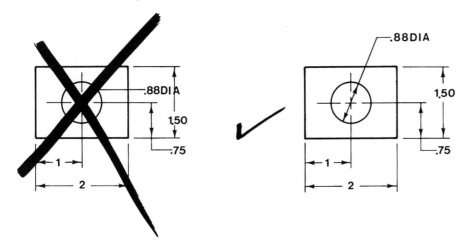

FIGURE 6-4(j)

(k) Use either decimals or fractions. Do not mix the two. Some companies make exception to this by having critical dimensions written in decimal form and noncritical dimensions written in fractional form. No such variance is in effect for the problems in this book.

FIGURE 6-4(k)

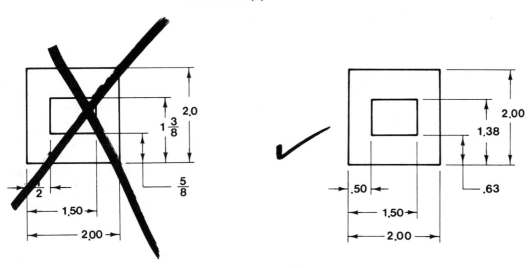

(l) Use either the unidirectional or aligned system. Do not mix the two.

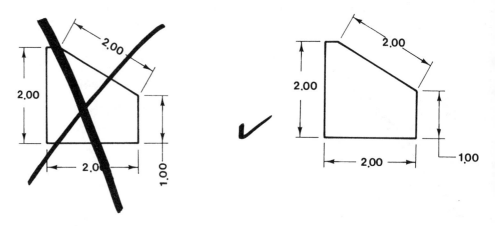

FIGURE 6-4(l)

(m) Unless it is absolutely necessary, do not dimension to a hidden line. In most cases, the addition of a section cut (see Chapter 10) to the drawing is probably the best way to eliminate excess or confusing hidden lines.

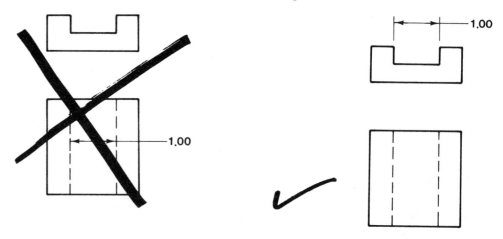

FIGURE 6-4(m)

(n) Do not include the symbol for inches (″) on dimensions. All dimensions in mechanical drafting are in inches unless otherwise stated. An exception to this rule is the number 1, which is usually written 1″ so that it will not make vertical dimensions lines appear as centerlines.

FIGURE 6-4(n)

(o) Always specify whether a hole or arc dimension is a diameter or a radius. Usually holes are dimensioned in diameters (DIA) and arcs are dimensioned by radii (R).

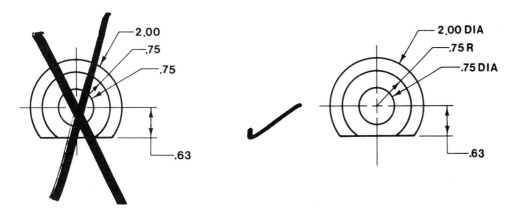

FIGURE 6-4(o)

(p) Do not run extension or dimension lines through other dimension or extension lines unless there is absolutely no alternative. The same is true for leader lines.

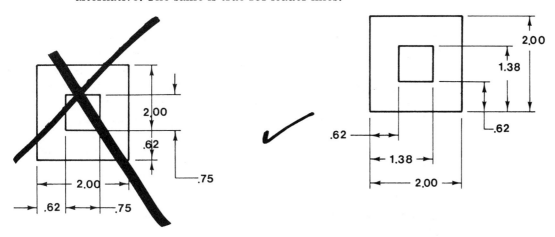

FIGURE 6-4(p)

(q) Always include overall dimensions except on objects that have rounded ends. This means the total length, width, and height for rectangular objects and the largest outside diameter and height for cylinders.

FIGURE 6-4(q)

(r) Always dimension holes in the views in which they appear as circles.

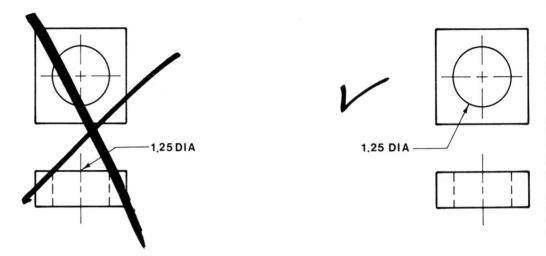

1.25 DIA

1.25 DIA

FIGURE 6-4(r)

6-4 UNIDIRECTIONAL AND ALIGNED SYSTEMS

Dimensions may be positioned on a drawing by using either the unidirectional or the aligned system. The unidirectional system is the preferred system. In the unidirectional system, all dimensions are placed so that they may be read from the bottom of the drawing, that is, with their guidelines horizontal. In the aligned system, dimensions are placed so that they may be read from either the bottom or the right side of the drawing, that is, with their guidelines parallel to the surface that they are defining. Figure 6-5 illustrates the difference between the two systems.

The unidirectional system is the newer of the two systems and it has become the most popular because it is easier to draw and to read. All problems in this book are dimensioned by using the unidirectional system.

FIGURE 6-5 Comparison between the unidirectional and aligned dimensioning system.

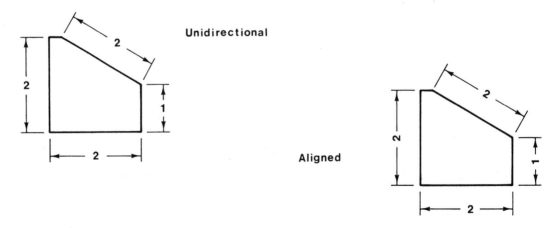

Unidirectional

Aligned

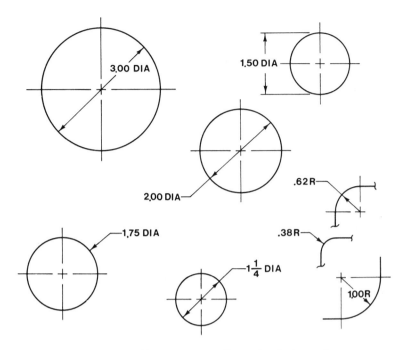

FIGURE 6-6   Different ways to dimension holes and arcs.

6-5   DIMENSIONING HOLES

Figure 6-6 illustrates several different ways to dimension holes. Holes are usually dimensioned to their diameters because most drills, punches, and boring machines are set up in terms of diameters. Arcs are usually dimensioned according to their radii.

Always locate a hole by dimensioning to its center point. Make sure that the center point of the hole is clearly defined by crossing the short sections of centerlines. The long sections of the centerlines may be dimensioned as if they were extension lines.

When you use leader lines, always point the arrow end of the line at the center point of the hole. Always finish the nonarrow end with a short horizontal section that will guide the reader's eye into the dimension note. Always place dimension notes so that they may be read from the bottom of the drawing.

Figure 6-7 shows two ways to dimension holes that do not pass

FIGURE 6-7   How to dimension hole depth.

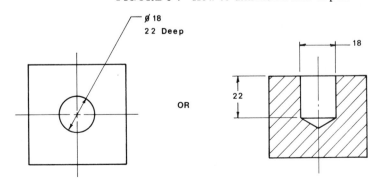

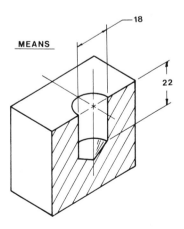

completely through an object. The first method uses a note consisting of the hole diameter followed by the required depth. The dimensional leader line should be applied to the circular view of the hole (where it appears as a circle).

The second method is applied to a side sectional view of a hole. The diameter and depth are referenced directly to the hole. Note that the 30° conical bottom portion of the hole is *not* considered as part of the depth measurement.

6-6 DIMENSIONING ANGLES

Figure 6-8 shows four different ways to dimension angles. In each method, the circular dimension line is drawn using a compass. The compass point is located at the vertex of the angle.

FIGURE 6-8 How to dimension features using angular dimensions.

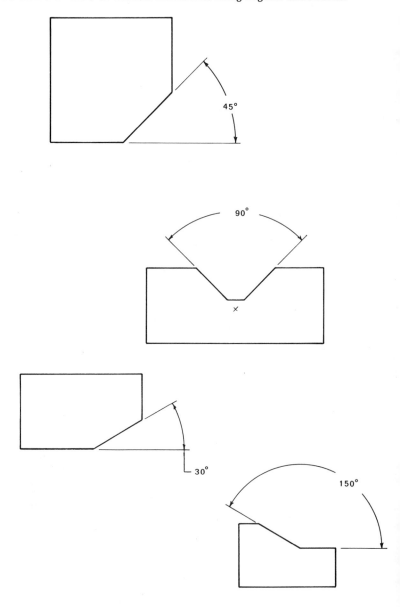

6-7   CIRCULAR HOLE PATTERNS

Figure 6-9 shows four examples of how to dimension hole patterns. In each example, the holes are located along a common diameter and by an angular dimension. Holes located on the vertical or horizontal center-lines do not need angular locational dimensions.

FIGURE 6-9   How to dimension multiple hole patterns.

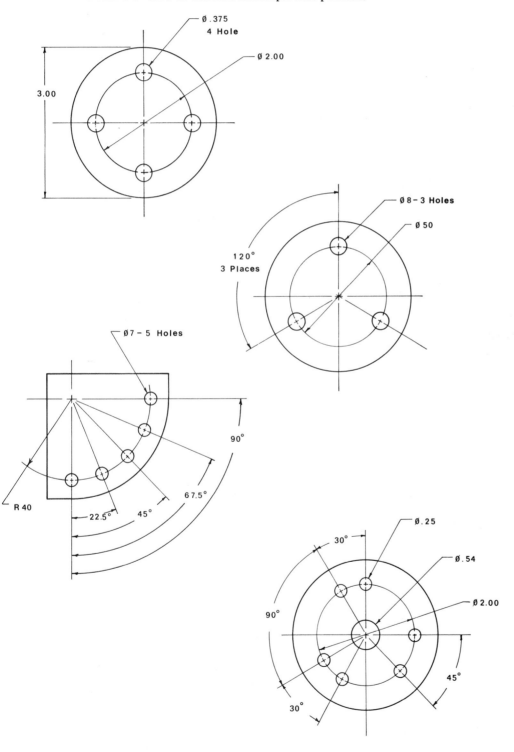

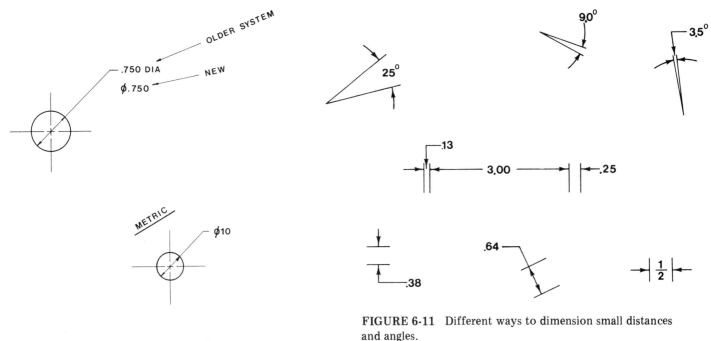

FIGURE 6-11 Different ways to dimension small distances and angles.

FIGURE 6-10 New symbol for diameter.

6-8 NEW DIAMETER SYMBOL

The latest revision of the ANSI Y14.5-1982 recommends that the notation DIA used to reference a diameter be replaced by the symbol ϕ. This symbol is placed in front of the numerical value as shown in Figure 6-10.

6-9 DIMENSIONING SMALL DISTANCES AND SMALL ANGLES

When you dimension a small distance or a small angle, always keep the lettering at the normal height of either 1/8 or 3/16. The temptation is to squeeze the dimensions into the smaller space. This is unacceptable because crowded or cramped dimensions are difficult to read, especially on blueprints which are microfilmed. Figure 6-11 shows several different ways to dimension small distances or angles and still keep the dimensions at the normal height.

6-10 BASELINE SYSTEM

The baseline system of dimensioning is illustrated in Figure 6-12. All dimensions in the same plane are located from the same line, which is called a *baseline*. (It is sometimes called a *reference line* or a *datum line*.) This system is particularly useful because it eliminates tolerance buildup, it is easy for manufacturers and inspectors to follow, and it is easily adaptable to the requirements of numerical tape machines. Its chief disadvantage is that the amount of space used on the drawing paper is larger—usually at least twice the area of the surface being defined. Also, once it is set up, it is difficult to alter.

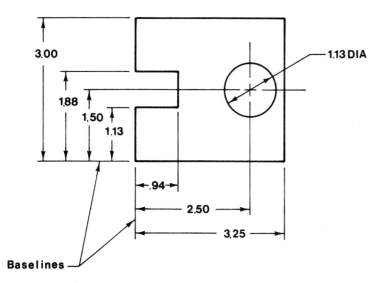

FIGURE 6-12 Baseline system.

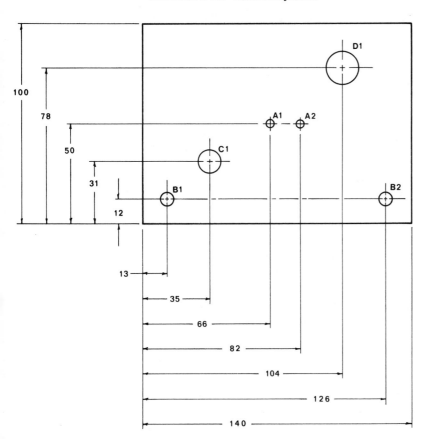

SIZE SYMBOL	A	B	C	D
HOLE DIA	4	7	11	17

FIGURE 6-13 Baseline system of dimensions.

When you use the baseline system, be careful to include all needed dimensions and be sure to use a large enough piece of paper.

Objects that contain different-sized holes may use a chart to specify the hole diameters, as shown in Figure 6-13. Each hole is assigned a letter equivalent to its diameter as defined in the chart. Each hole is also assigned a number for referencing purposes.

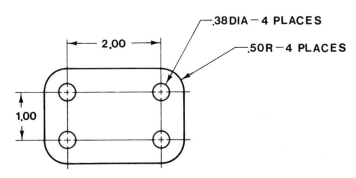

FIGURE 6-14   Hole-to-hole system.

6-11   HOLE-TO-HOLE SYSTEM

The hole-to-hole system is illustrated in Figure 6-14. It is a modification of the baseline system (Section 6-10) which is used to dimension parts whose hole-to-hole distances are critical, for example, a part that must align with the shafts or dowels of another part for proper assembly.

In the hole-to-hole system, all dimensions in the same plane are measured for the lines that define the critical holes. The baseline is not, in this case, a physical line, but it is the centerline between the critical holes.

6-12   COORDINATE SYSTEM

The coordinate system is a dimensioning system based on the mathematical x-y coordinate system. It is usually only used to dimension an object that contains a great many holes, for example, an electrical chassis. It is particularly well suited to computer use and numerically controlled tape machines.

Each hole on the given surface is located relative to an x-y coordinate system and then all values are listed in a chart. The overall dimensions are not included in the chart but are located on the picture part of the drawing. Figure 6-15 is an example of an object dimensioned by using the coordinate system.

FIGURE 6-15   Coordinate system.

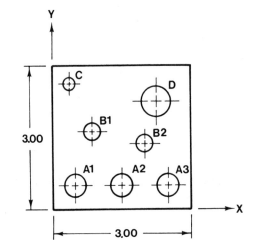

HOLE	X	Y	DIA
A1	0.50	0.50	.470
A2	1.50	0.50	.470
A3	2.50	0.50	.470
B1	0.88	1.62	.375
B2	2.00	1.38	.375
C	0.38	2.62	.250
D	2.25	2.25	.625

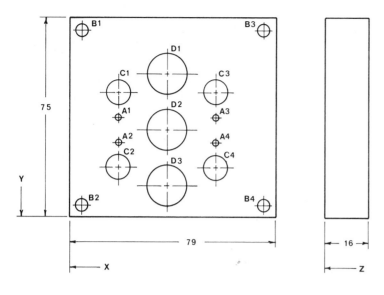

HOLE	X	Y	Z	Ø	
A1	18	37	THRU	2.5	
A2	18	28	↓	↓	
A3	56	37	↑	↑	
A4	56	28	THRU	2.5	
B1	4	70	12	4.5	
B2	4	4	12	↓	
B3	75	70	12	↑	
B4	75	4	12	4.5	
C1	19	47	THRU	8	
C2	19	19	↓	↓	
C3	56	47		↑	
C4	56	19		8	
D1	37	54		15	
D2	37	33	↑	15	
D3	37	12	THRU	15	

FIGURE 6-16 Coordinate system of dimensions.

Figure 6-16 shows a three-dimensional coordinate system. Hidden lines are omitted from the side view. Hole depths are listed in the Z column. Holes that go completely through the object are listed as THRU in the Z column.

6-13 BASELINE SYSTEM WITHOUT DIMENSION LINES

Figure 6-17 shows an object dimensioned using the baseline system without dimension lines. This system eliminates the difficulty in drawing the many dimensions and extension lines required by the baseline system.

The number located at the end of each hole centerline specifies the distance of the centerline from the 0, 0 point. Hole sizes are listed in a table as shown.

FIGURE 6-17 Baseline system without dimension lines.

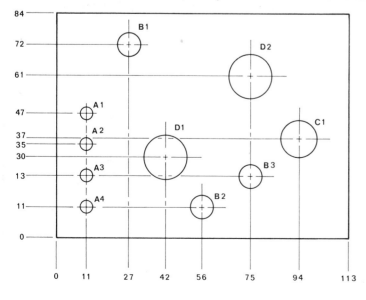

SIZE SYMBOL	A	B	C	D
HOLE DIA	5	9	14	16

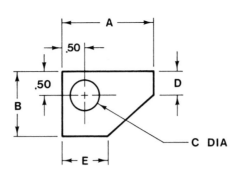

PART NO	A	B	C	D	E
1001	2.00	1.38	.68	.50	1.00
1002	2.00	1.38	.50	.68	1.13
1003	2.25	1.50	.68	.75	1.13
1004	2.25	1.50	.50	.75	1.25
1005	2.38	1.75	.68	.75	1.25

FIGURE 6-18 Tabular dimensions.

6-14 TABULAR DIMENSIONS

Often, manufacturers will produce a part in several different sizes. Each part will have the same basic shape, but the part will vary in overall size. To save having to dimension each part individually, a system called *tabular dimensioning* is used. Figure 6-18 illustrates an example of tabular dimensioning.

To read tabular dimensions, look up the part number in the table and substitute the given numerical values for the appropriate letters in the figure. For example, part number 1003, according to the table, has an *A* value of 2.25, a *B* value of 1.50, and so on. Part number 1005 has an *A* value of 2.50, a *B* value of 1.75, and so on. The numerical dimensions of 0.50 mean that these dimensions do not vary, that they remain the same for all parts.

The table may also be used in reverse. If you know what your given design requirements are, look up these values in the table to find which part number you should call out on the drawing.

6-15 IRREGULARLY SHAPED CURVES

To dimension an irregularly shaped curve, dimension the points that define the line. The more points you dimension, the more accurate will be your definition. Figure 6-19 illustrates a dimensioned irregularly shaped curve.

FIGURE 6-19 Dimensioning an irregular curve.

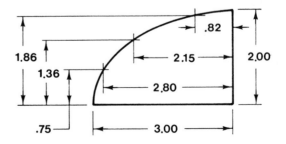

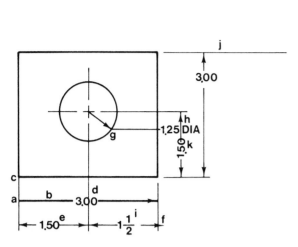

FIGURE 6-20   Some common dimensioning errors.

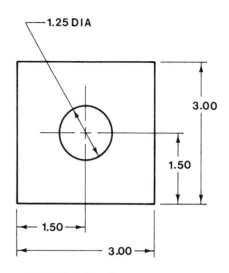

FIGURE 6-21   Errors illustrated in Figure 6-20 corrected.

6-16   COMMON DIMENSIONING ERRORS

Figure 6-20 demonstrates some of the most common dimensioning errors. Note how cluttered and confined the dimensions appear. Compare Figure 6-20 with Figure 6-21. Both drawings are of the same shape, but Figure 6-21 is dimensioned properly.

Study the errors in Figure 6-20 and then see how they were corrected in Figure 6-21. The errors are:

a. No arrowhead

b. Dimension line too thick

c. No gap between object and extension line

d. Dimension value placed over a centerline

e. No gap between dimension line and dimension value

f. Arrowhead extends beyond extension line

g. Leader line changes direction within the object

h. Dimension value written too close to the object and over a dimension line

i. Fraction used while all other dimensions are in decimal form

j. Extension line too long

k. Dimension not written horizontally

6-17   SLOTS

Figure 6-22 shows three ways of dimensioning slots. Note that no specific value for R is given, just the letter R. This differs from past practice, but is more compatible with modern manufacturing and inspection methods.

Figure 6-23 shows how the dimensioning systems shown in Figure 6-22 are applied to patterns of slots. Each system refers the slot pattern to a common datum surface.

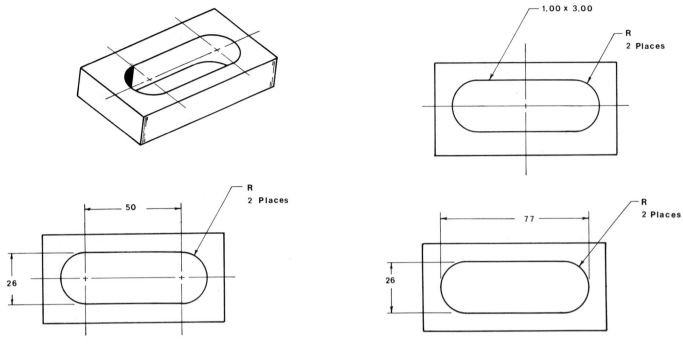

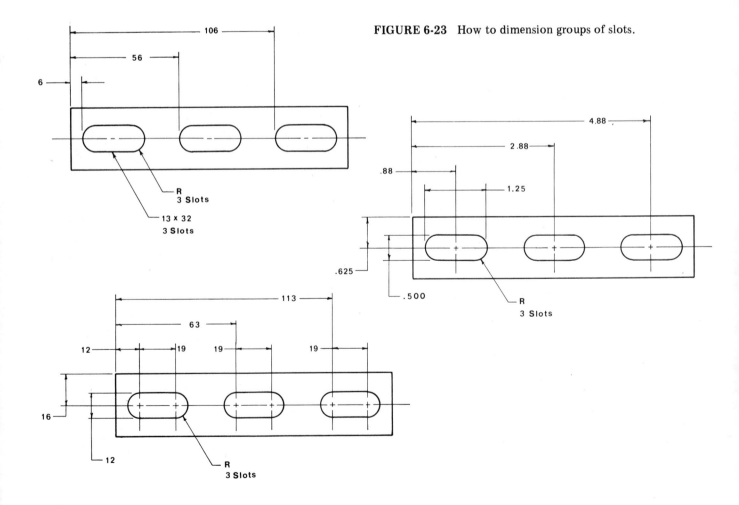

FIGURE 6-22   How to dimension slots.

FIGURE 6-23   How to dimension groups of slots.

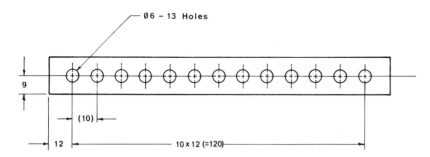

FIGURE 6-24 How to dimension evenly spaced holes.

6-18 HOLE PATTERNS

Figure 6-24 shows how to dimension a group of evenly spaced holes of equal diameter. The dimension enclosed in parentheses is a referenced dimension. This means that it is for reference only and not for inspection.

Hole patterns that include holes of different diameters, but that use the same circular centerline, are dimensioned as shown in Figure 6-25. If the difference in hole size is obvious, holes need to be identified only by numerical value. If the difference is not obvious, identify size by letter, as shown.

FIGURE 6-25 How to dimension different diameter holes on the same circular centerline.

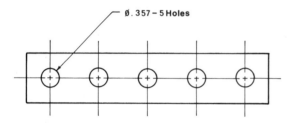

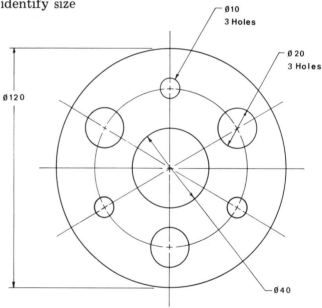

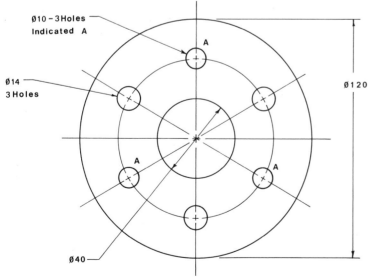

147

6-19 COUNTERSINK AND COUNTERBORE

Countersink holes are dimensioned as shown in Figure 6-26. The first diameter is the drill diameter followed by the drill depth. The second diameter is the countersink diameter. Countersinks are manufactured at 82° but are drawn at 90°.

Counterbored holes are dimensioned as shown in Figure 6-27. The first diameter is the drill diameter, the second the counterbored diameter.

FIGURE 6-26   How to dimension a countersunk hole.

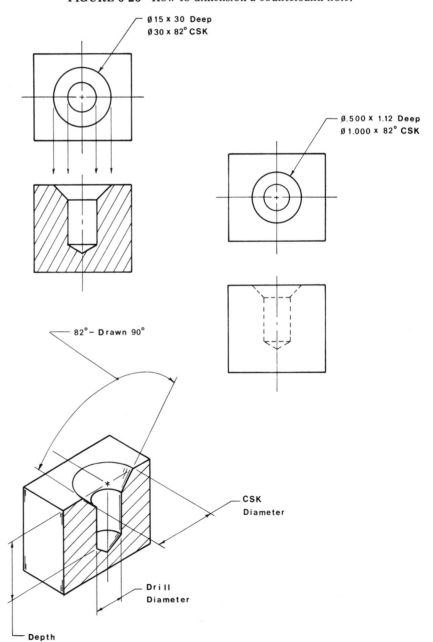

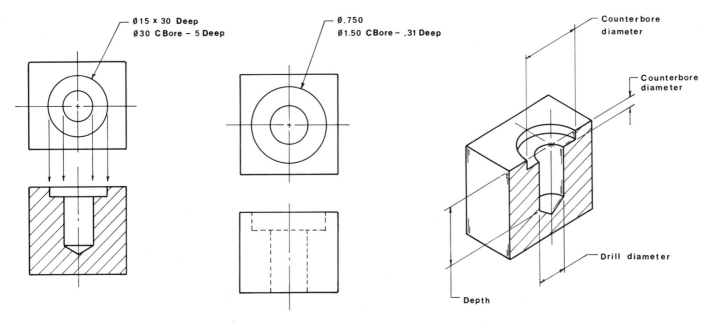

FIGURE 6-27 How to dimension a counterbored hole.

6-20   TOLERANCES

No dimension can be made perfectly. Unless you are very lucky, there will always be some variance. If, for example, you call for a dimension to be made 5 inches long, you will not get exactly 5 inches on the finished part. It may measure 5.0001 or 4.99999 or some other value, but it will not be exactly 5 inches.

It is not only impossible to manufacture perfect dimensions, it is also unnecessary. It is possible for a carpenter to build a house within the nearest 0.01 inch, but it is not necessary for the structural soundness of the house. Think of how much time such a constraint would add to the normal time required to build a house, and then think of how this extra time would needlessly increase the building cost of the house.

Because it is impossible to manufacture perfect dimensions, all dimensions must be toleranced. Each dimension must be considered separately in regard to how much variance is acceptable to ensure a satisfactory finished product. The final judgment must be made by considering, among other things, manufacturing capabilities, customer requirements, usage requirements, material properties, and cost constraints. It takes experience and practice to make such a judgment correctly.

Figure 6-28 shows four ways that tolerance can be added to dimensions on a drawing. Note that no zero is required to the left of the zero when English units (inches) are used, but a zero is required when metric (millimeters) units are used.

Bilateral tolerances define a tolerance range that is equally spaced around a base dimension. For example, in Figure 6-28 0.88 is the base dimension and +0.02 is the bilateral tolerance. This means that values as large as 0.90 (0.88 + 0.02) or as small as 0.86 (0.88 − 0.02) are acceptable.

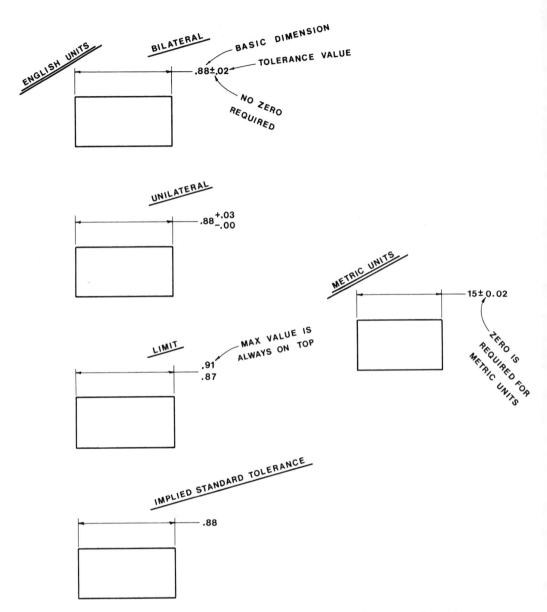

FIGURE 6-28 How to add tolerance callouts to a dimension.

Unilateral tolerances are applied in one direction from a base dimension. In Figure 6-28, the base dimension 0.88 can be as large as 0.91 (0.88 + 0.03) but no smaller than 0.88 (0.88 − 0.00) under the unilateral tolerances specified.

Limit tolerances simply state the maximum and minimum values of the dimensions. In Figure 6-28 the maximum value is 0.91 and the minimum 0.87. The maximum value is always written above the minimum value.

If no tolerance value is stated, it implies that a standard tolerance is to be used. Most companies have standard tolerances based on their product line. Standard tolerance values are usually printed on the drawing paper next to the title block.

Figure 6-29 shows two typical standard tolerance callout blocks, one for English units, the other for metric units. The XX notation used means any number, so 0.XX would mean any dimension using two

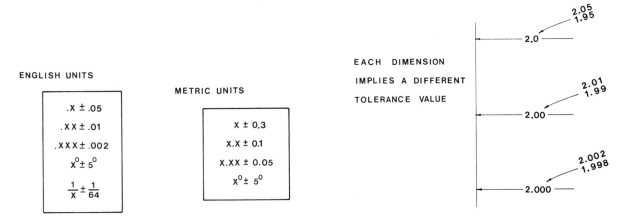

FIGURE 6-29   A sample standard tolerance.   FIGURE 6-30   How to interpret implied tolerances.

decimal places (0.32, 2.25, 0.04, 0.20). Any dimension using two decimal places has an implied standard tolerance value of +0.01 if the values defined in Figure 6-30 are used.

Numbers that are equal mathematically may have different tolerance values. In Figure 6-30, the dimensions 2.0, 2.00, and 2.000 are equal from a mathematical standpoint, but are very different from a manufacturing standpoint. The 2.0 dimension would have a tolerance of +0.05 (see Figure 6-29), whereas the 2.000 dimension would have a tolerance of +0.002.

6-21   CUMULATIVE TOLERANCES

Cumulative tolerances are errors that occur when several small, seemingly insignificant errors are compounded. Usually, they are the result of improper dimensioning. For example, consider Figure 6-31(a), and assume that the object is being manufactured to a standard company tolerance of ±0.02 for all two-place dimensions. Each of the 1.00 dimensions could be made 1.02, giving an overall length of 4.08.

$$\begin{array}{r} 1.02 \\ 1.02 \\ 1.02 \\ \underline{1.02} \\ 4.08 \end{array}$$

4.08 is not an acceptable overall length, since the overall length must be, according to the given dimension, 4.00 ± 0.02. This means that the greatest acceptable length is 4.02. The 4.08 object would not pass inspection. Unfortunately, the responsibility for this error must be placed directly on the drafter who improperly dimensioned the object.

There are several other ways to avoid the error in Figure 6-31(a). In Figure 6-31(b), a REF note (reference note) was placed on the 4.00 dimension. The REF notation means that the dimension is not critical to the manufacturing of the object and has only been included for the reader's convenience. Be very careful when you use the REF notation that it is used only for insignificant dimensions.

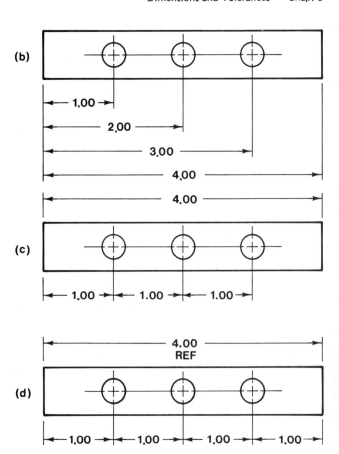

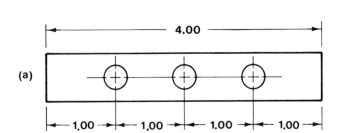

FIGURE 6-31   Cumulative tolerances.

In Figure 6-31(c) one of the 1.00 dimensions was dropped. This means that the end section of the object, which is now not dimensioned, may vary. All the other dimensions are still manufactured within stated tolerances. In Figure 6-31(d) the baseline system of dimensioning was used. In most cases, the baseline system is the best way to avoid cumulative tolerances because no one dimension is dependent on the accuracy of another dimension. Each dimension is manufactured separately.

6-22   TOLERANCE STUDIES

A tolerance study is an analysis of the range of values created by tolerances, either individually or in groups. For example, in Figure 6-32, given the two dimensions and tolerances, determine the maximum and minimum values possible for the overall length. The maximum overall length is found by adding the maximum values of the two dimensions $(1.52 + 1.53 = 3.05)$. The minimum overall length is found by adding the minimum values of the two dimensions $(1.48 + 1.47 = 2.95)$.

In Figure 6-33, determine the maximum and minimum values for surface A. Surface A will be at its maximum when the overall length is at maximum value and when both surfaces B and C are at their minimum values $(3.02 - 0.99 - 0.97 = 1.06)$. Surface A will be at its minimum when the overall length is at its minimum and surfaces B and C are at their maximum $(2.98 - 1.01 - 1.03 = 0.94)$.

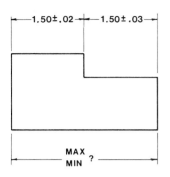

MAX	MIN
1.52	1.48
1.53	1.47
3.05	2.95

FIGURE 6-32 A sample tolerance study.

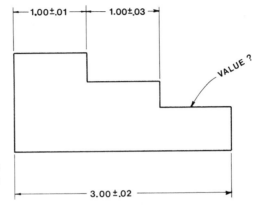

MAX	MIN
3.02	2.98
-.99	-1.01
-.97	-1.03
1.06	.94

FIGURE 6-33 How to determine the maximum and minimum value of a fracture.

PROBLEMS

P6-1 through P6-6   Redraw and dimension each shape or object. Each square on the grid pattern is 1/4 per side.

FIGURE P6-1

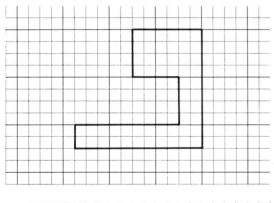

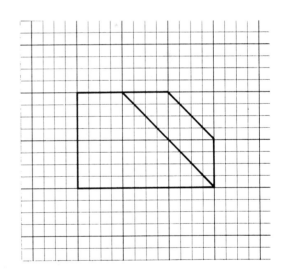

FIGURE P6-3

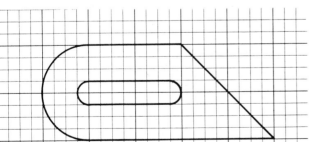

FIGURE P6-2

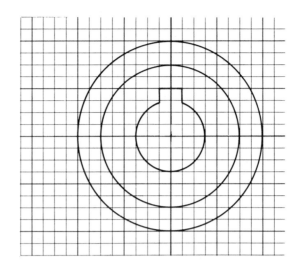

FIGURE P6-4

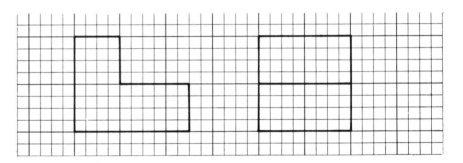

FIGURE P6-5

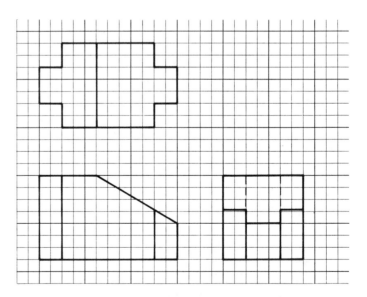

FIGURE P6-6

P6-7 Redraw and dimension each shape. Each square on the grid
through pattern is 0.20 per side.
P6-17

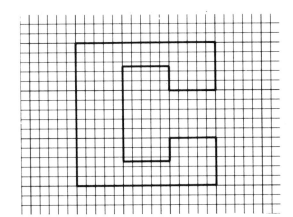

FIGURE P6-7

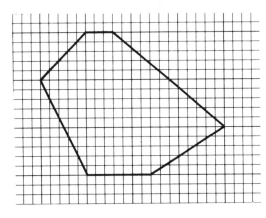

FIGURE P6-9

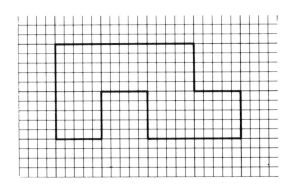

FIGURE P6-8

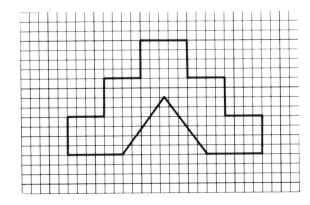

FIGURE P6-10

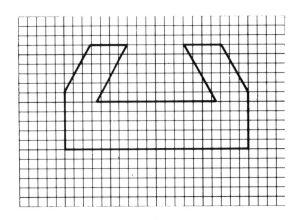

FIGURE P6-11

FIGURE P6-13

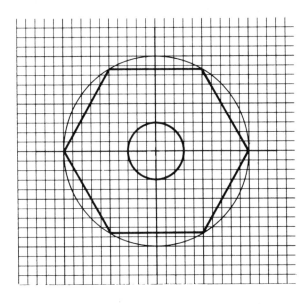

FIGURE P6-12

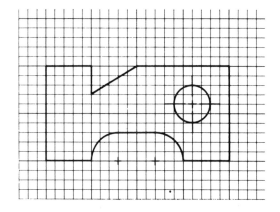

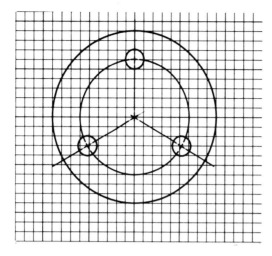

FIGURE P6-14

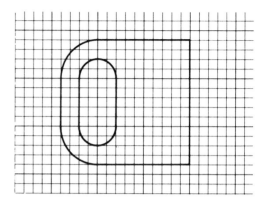

FIGURE P6-15

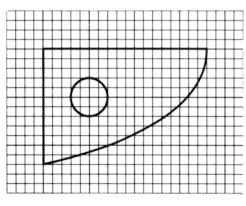

FIGURE P6-16

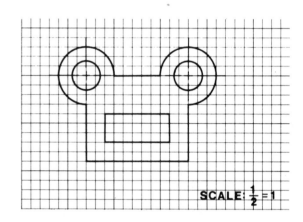

SCALE: $\frac{1}{2}$ = 1

FIGURE P6-17

P6-18   Dimension the chassis surface shown in Figure P6-18 twice, once using the baseline system and once using the coordinate system.

FIGURE P6-18

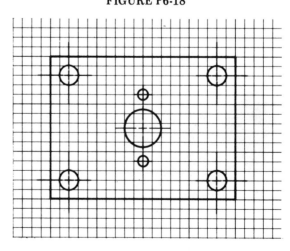

P6-19 Draw three views of each object and dimension. Use both the
through decimal and the unidirectional system. Each triangle or square
P6-24 on the grid pattern is 0.20 per side.

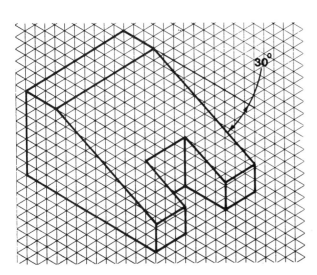

FIGURE P6-19

FIGURE P6-20

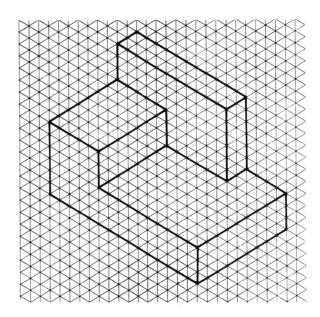

FIGURE P6-21

FIGURE P6-22

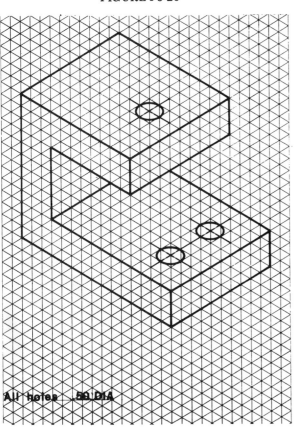

All holes .50 DIA

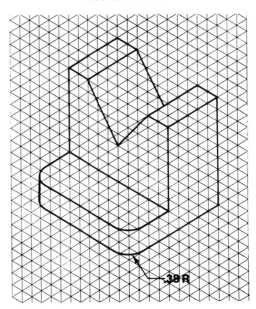

.38 R

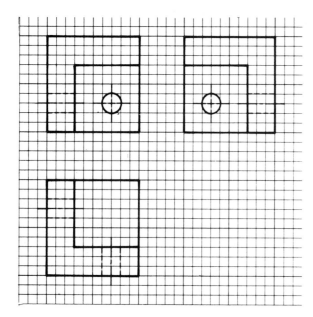

FIGURE P6-23

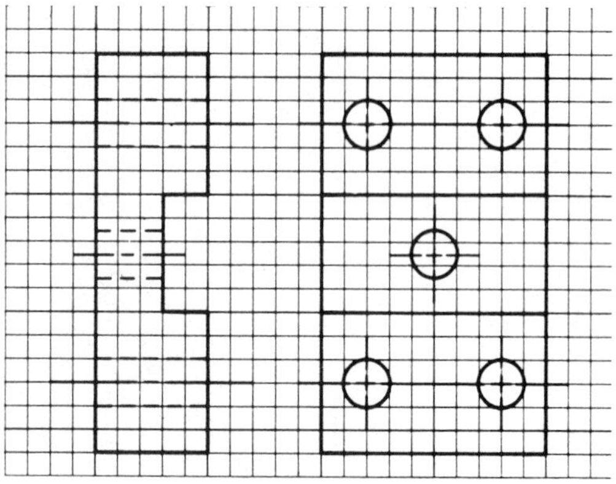

FIGURE P6-24

P6-25 What is the maximum and minimum height that the five pieces shown in Figure P6-25 could generate if their dimensions and tolerances are as follows:

a. 1.38 ± 0.06

b. $0.63 \begin{array}{l} +0.00 \\ -0.05 \end{array}$

c. $1.50 \begin{array}{l} +0.02 \\ -0.00 \end{array}$

d. 1.000 ± 0.004

e. $1\text{-}3/8 \pm 1/32$

FIGURE P6-25

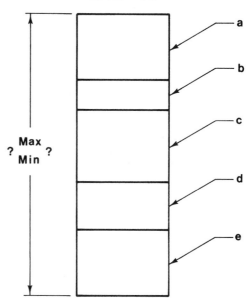

P6-26   In Figure P6-26 all linear dimensions have a tolerance of ±0.03 and all angular dimensions have a tolerance of ±1°. If the disk piece is placed within the 90° opening of the larger base piece, what is the maximum height that the two pieces together could generate? Prepare a layout to verify your answer.

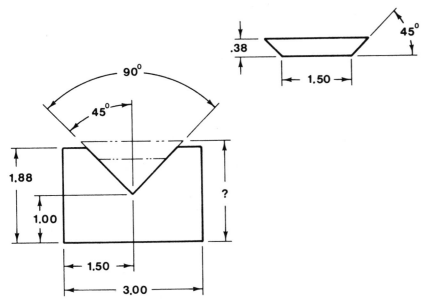

FIGURE P6-26

P6-27   Redraw the object shown in Figure P6-27 and insert the following dimensions.

a. $2.75 \begin{smallmatrix} +0.01 \\ -0.02 \end{smallmatrix}$

b. 1.88 ± 0.01

c. 1.130 ± 0.002

d. 2.38 ± 0.01

e. $2.000 \begin{smallmatrix} +0.003 \\ -0.002 \end{smallmatrix}$

f. $1.38R \begin{smallmatrix} +0.00 \\ -0.01 \end{smallmatrix}$

g. $2.00 \begin{smallmatrix} +0.05 \\ -0.00 \end{smallmatrix}$

h. $30° \pm 5°$

i. 0.750 ± 0.001

j. 1.25 ± 0.03

FIGURE P6-27

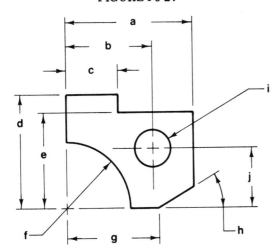

P6-28 Dimension the object shown in Figure P6-28 by using the tabular system. For part number

1001: $V_1 = 3.00, V_2 = 1.80, V_3 = 2.00, V_4 = 0.80$
1002: $V_1 = 3.20, V_2 = 1.90, V_3 = 2.00, V_4 = 0.80$
1003: $V_1 = 4.00, V_2 = 2.20, V_3 = 2.20, V_4 = 0.80$
1004: $V_1 = 4.00, V_2 = 2.20, V_3 = 2.40, V_4 = 1.00$

Each square on the grid pattern is 0.20 per side.

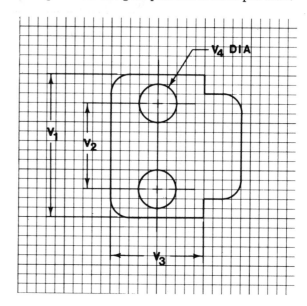

FIGURE P6-28

P6-29 Redraw Figure P6-29 using the dimensions and tolerances listed below. Add the dimensions and tolerances in the appropriate places. All values are in inches.

1. $1.50 \pm .02$
2. $1.50 \pm .03$
3. $.63 \pm .01$
4. $.75 \pm .02$

5. $.63 \pm .01$
6. $2.25 \pm .01$
7. $.500 \pm .001$

FIGURE P6-29

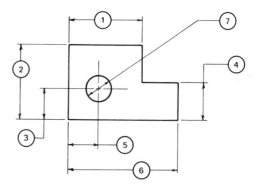

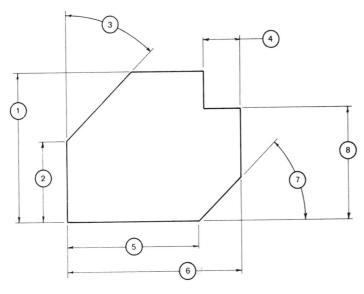

FIGURE P6-30

P6-30 Redraw Figure P6-30 using the dimensions and tolerances listed below. Add the dimensions and tolerances in the appropriate places. All values are in inches.

1. $\begin{array}{c} 3.002 \\ 2.999 \end{array}$

2. $1.630 \begin{array}{c} + .001 \\ - .003 \end{array}$

3. $45° \pm 3°$

4. $.75 \pm .02$

5. $\begin{array}{c} 2.754 \\ 2.749 \end{array}$

6. $3.50 \pm .01$

7. $45.0° \pm 0.5°$

8. $2.250 \begin{array}{c} + .000 \\ - .005 \end{array}$

P6-31 Redraw Figure P6-31 using the dimensions and tolerances listed below. Add the dimensions and tolerances in the appropriate places. All values are in millimeters.

1. 34 ± 0.1
2. 17 ± 0.05
3. 25 ± 0.03
4. 15 ± 0.5
5. 50 ± 0.03
6. 80 ± 1.0

7. $R5 \pm 0.1$
   ALL AROUND
8. 45 ± 0.5
9. 60 ± 0.5
10. $\phi \begin{array}{c} 14.02 \\ 13.99 \end{array}$ – 3 HOLES
11. 15 ± 0.1
12. 30 ± 0.1

FIGURE P6-31

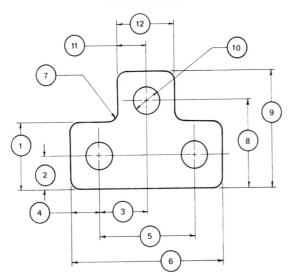

P6-32 Redraw Figure P6-32 using the dimensions and tolerances listed below. Add the dimensions and tolerances in the appropriate places. All values are in inches.

1. 2.00 ± 0.02
2. $R1.75 \pm 0.05$
3. 2.50 ± 0.01
4. $\dfrac{3.02}{3.00}$
5. $\dfrac{1.51}{1.49}$
6. $\phi^{0.500}_{0.497}$ – 3 HOLES
7. 1.250 ± 0.001
8. 1.250 ± 0.001
9. 4.000 ± 0.04

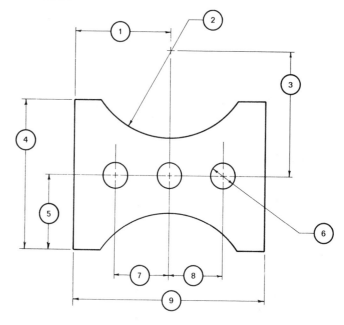

FIGURE P6-32

P6-33 Redraw Figure P6-33 using the dimensions and tolerances listed below. Add the dimensions and tolerances in the appropriate places. All values are in inches.

1. $R2.00 \pm .05$
2. $\dfrac{3.002}{2.999}$
3. $\dfrac{1.501}{1.498}$
4. $1.250 \pm .001$
5. $2.875 \pm .002$
6. $2.000 \pm .002$
7. $45.0 \pm 0.2° -$ BOTH SIDES
8. $1.00 \pm .01$
9. $1.50 \pm .01$
10. $1.000 \pm .002$
11. $\dfrac{2.003}{1.999}$
12. $0.750 \pm .005$
13. $7.00 \pm .01$

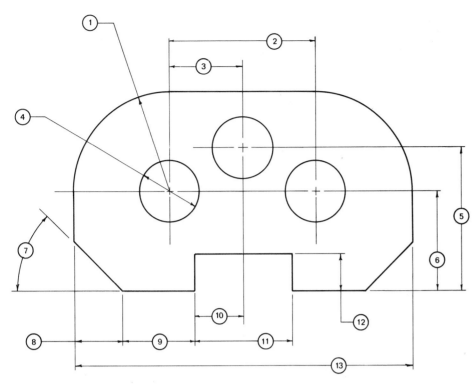

FIGURE P6-33

P6-34   Redraw Figure P6-34 using the dimensions and tolerances listed below. Add the dimensions and tolerances in the appropriate places. All values are in millimeters.

1. $\phi 30.0 \pm 0.01$

2. $\phi^{15.01}_{14.99}$

3. 10

4. 20

5. $67.0 ^{+\,0}_{-\,0.2}$

6. 15 ± 0.02

7. 35

8. 70

9. ALL FILLETS AND ROUNDS = R5.0

FIGURE P6-34

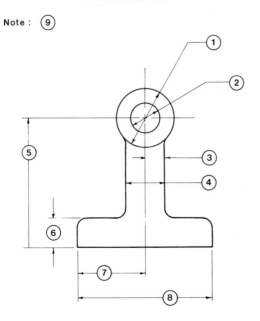

P6-35
through
P6-54
Each shape is shown at scale: $1/2 = 1$. Redraw the shape full size by measuring each drawing and multiplying the distances by 2. Add appropriate dimensions. Figures may be measured using either inches or millimeters. If inches are used, measure to the nearest 1/32 or .03 inch. If millimeters are used, measure to the nearest millimeter.

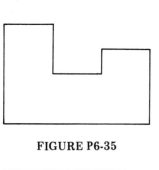

FIGURE P6-35

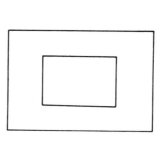

FIGURE P6-36

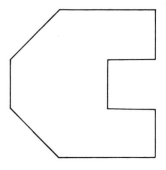

FIGURE P6-37

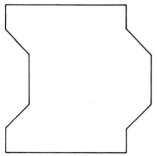

FIGURE P6-38

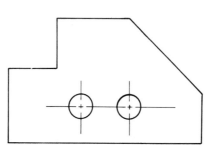

FIGURE P6-39

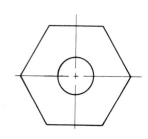

FIGURE P6-40

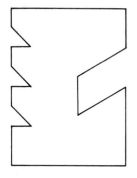

FIGURE P6-41

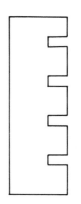

FIGURE P6-42

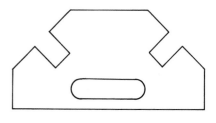

FIGURE P6-43

FIGURE P6-46

FIGURE P6-44

FIGURE P6-45

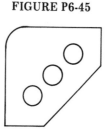

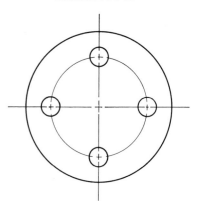

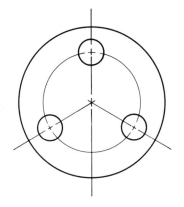

FIGURE P6-47

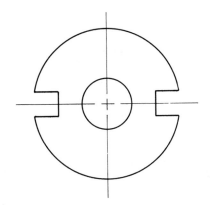

FIGURE P6-48

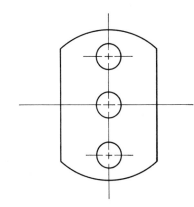

FIGURE P6-49

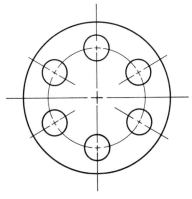

FIGURE P6-50

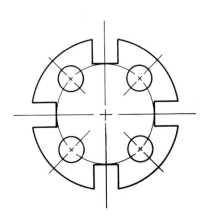

FIGURE P6-51

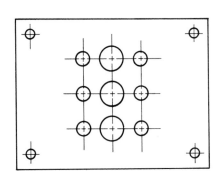

FIGURE P6-52

FIGURE P6-53

FIGURE P6-54

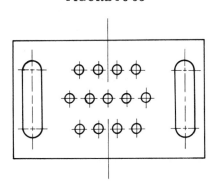

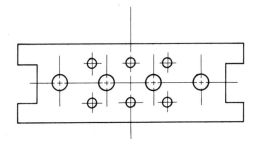

7

OBLIQUE SURFACES AND EDGES

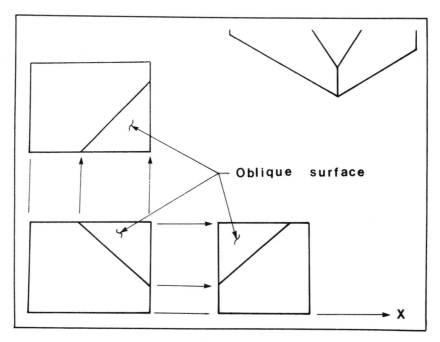

Oblique surface

FIGURE 7-0

7-1 INTRODUCTION

Oblique surfaces and edges are made up of planes and lines that are not parallel to either principal plane line. Figure 7-1 is an example of an oblique surface. Note that none of the lines that define the surface is parallel to either principal plane line and that each line is a different length in each given orthographic view. Note also that the shape of the plane also varies in each orthographic view. This variance makes it difficult to visualize what oblique surfaces really look like (what is their true shape) and will force you to rely on projection theory to help you to formulate accurate finished drawings.

This chapter explains and illustrates the kinds of oblique surfaces most often found in drawings.

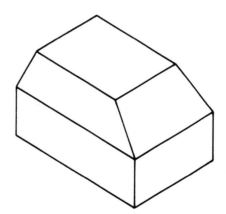

FIGURE 7-1 An oblique surface.

7-2 COMPOUND EDGES AND LINES

Figure 7-2 is a problem that involves a compound edge. The problem is to draw the top view, given the front and right-side views. Figure 7-3 is the solution and was derived by the following procedure:

GIVEN: Front and right-side views.
PROBLEM: Draw the top view.

FIGURE 7-2

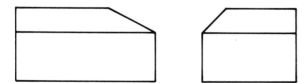

SOLUTION:

1. Make, to the best of your ability, a freehand sketch of the solution and, if possible, an isometric sketch of the entire object.

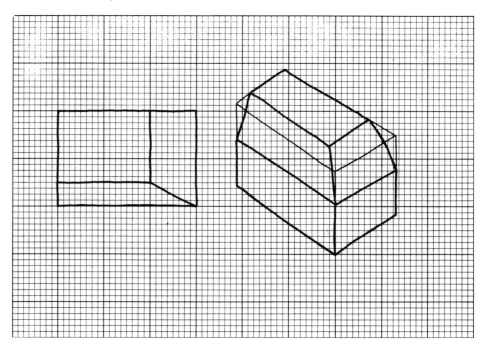

FIGURE 7-3(a)

2. Analyze the given information and label those points about which you are unsure. In this example, surfaces 1-2-3-4 and 3-4-5-6 were labeled.

FIGURE 7-3(b)

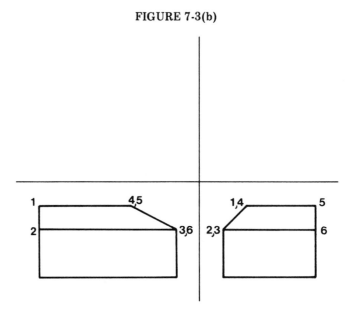

3. Project points 1, 2, 3 4, 5, and 6 into the top view by using the projection theory presented in Chapter 4.

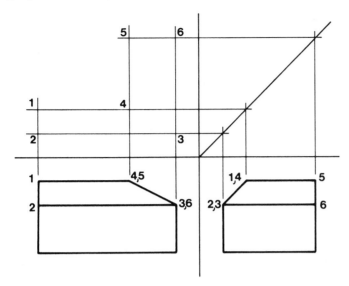

FIGURE 7-3(c)

4. Using very light construction lines, lay out the top view of surfaces 1-2-3-4 and 3-4-5-6. Also lightly lay out the remainder of the object.

FIGURE 7-3(d)

5. Erase all excess lines and darken in all the lines to their final color and configuration.

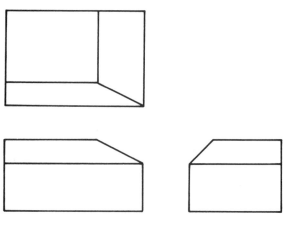

FIGURE 7-3(e)

In this example, line 3-4 is a compound edge. It was formed by the intersection of two inclined surfaces; yet line 3-4 is not parallel to either principal plane line.

Figure 7-4 is another problem that involves a compound edge. In this problem the object is pictured (an isometric drawing is presented) and you are asked to draw all three orthographic views: front, top, and right side. Figure 7-5 is the solution and was derived by the following procedure:

GIVEN: An object.
PROBLEM: Draw front, top, and side views.

SOLUTION:

1. Make, to the best of your ability, a sketch of the solution. Make the sketches as complete and as accurate as you can. It is much easier to change sketches than to change drawings.

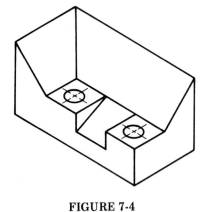

FIGURE 7-4

FIGURE 7-5(a)

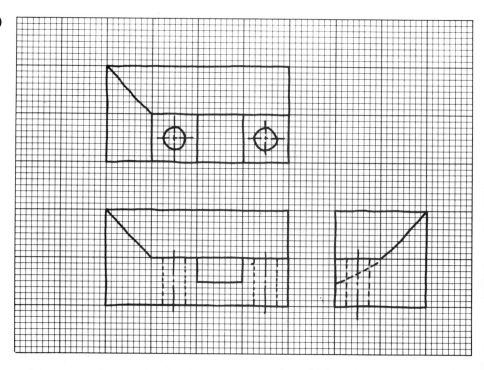

2. Working from your sketches, lightly lay out the solution. If necessary, label any confusing areas and use projection theory to work known pieces of information together to formulate the final solution. Also, use projection theory to check any areas about which you are unsure.

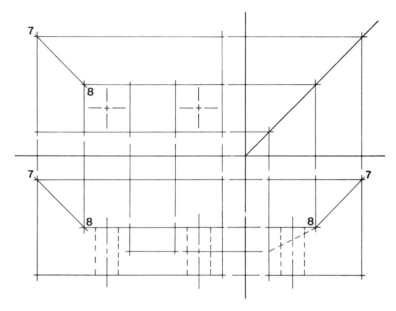

FIGURE 7-5(b)

3. When the layout is complete, erase all excess lines and draw in all lines to their final color and configuration.

FIGURE 7-5(c)

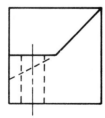

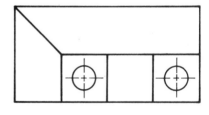

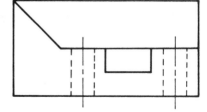

7-3   OBLIQUE SURFACES

Figure 7-6 is a problem that involves an oblique surface. An oblique surface is one that is not parallel to either principal plane line (see Figure 7-1). The problem is to draw the front view given the top and right-side views. Figure 7-7 is the solution and was derived by the following procedure:

GIVEN: Top and side views.
PROBLEMS: Draw the front view.

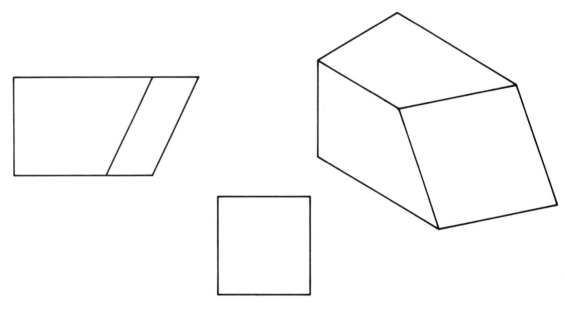

FIGURE 7-6

SOLUTION:

1. Make, to the best of your ability, a freehand sketch of the solution and, if possible, an isometric sketch of the entire project.

FIGURE 7-7(a)

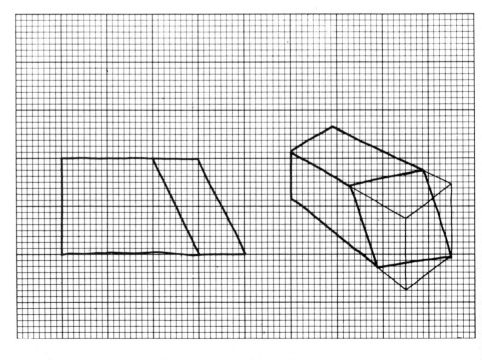

2. Analyze the given information and label those points, lines, or planes about which you are unsure. In this example surface 1-2-3-4 was labeled.

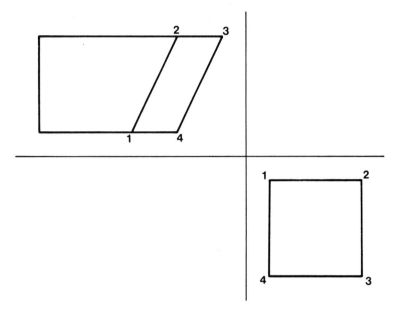

FIGURE 7-7(b)

3. Project points 1, 2, 3, and 4 into the front view by using projection theory.

FIGURE 7-7(c)

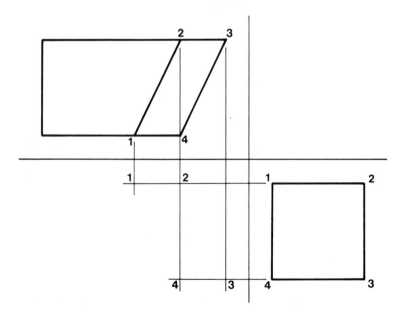

4. Using very light layout lines, lay out the front view of surface 1-2-3-4. Also lay out the remainder of the object.

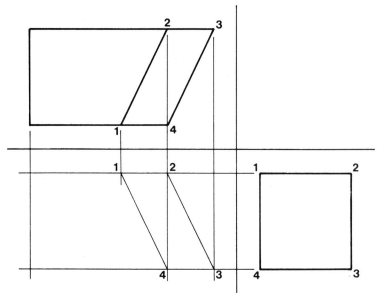

FIGURE 7-7(d)

5. Erase all excess lines and draw in all lines to their final color and configuration.

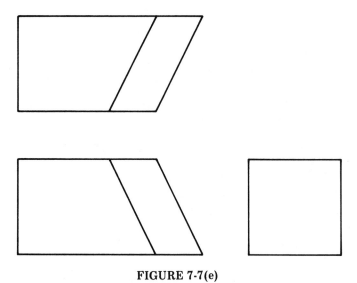

FIGURE 7-7(e)

Surface 1-2-3-4 in Figure 7-7 is an oblique surface. It is not parallel to either of the principal plane lines. Because it is not parallel to either principal plane line, none of the three final views represents a true picture of the shape of surface 1-2-3-4. How to find the true shape of an oblique surface is explained in Section 11-6.

Figure 7-8 is another example of a problem involving an oblique surface.

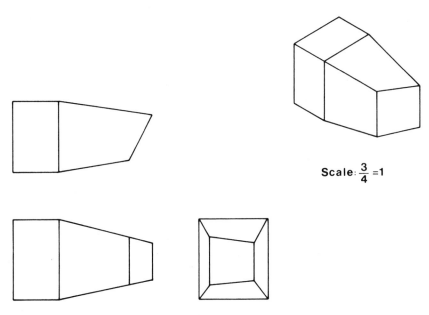

Scale: $\frac{3}{4}$ =1

FIGURE 7-8   Three views of an object that contains oblique surfaces.

7-4 PARALLEL EDGES

Figure 7-9 is an example of a problem that involves parallel edges. Parallel edges are edges that are parallel to each other and may or may not be parallel to the principal plane lines. The problem is to draw the front, top, and right-side views when an isometric drawing is given. Figure 7-10 is the solution and was derived by using the same procedure outlined for Figure 7-6.

GIVEN: An object.
PROBLEM: Draw front, top, and right-side views.

SOLUTION:

FIGURE 7-9

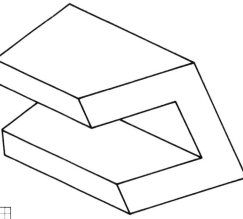

FIGURE 7-10(a)

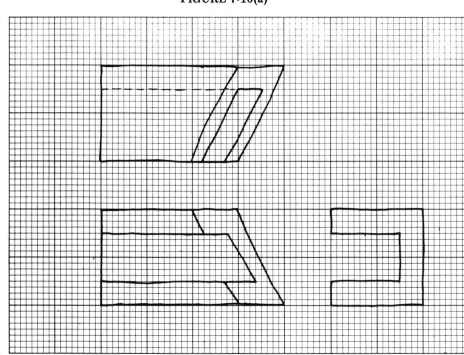

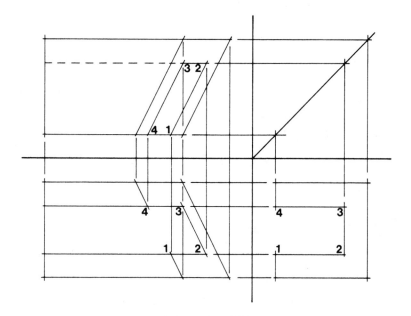

FIGURE 7-10(b)

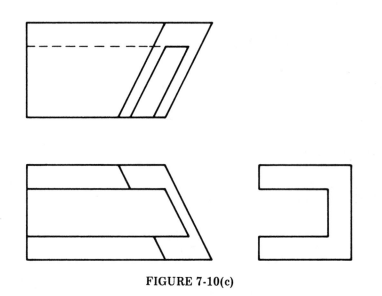

FIGURE 7-10(c)

In this problem surface 1-2-3-4-5-6-7-8 is an oblique flat surface that cuts across the object. The object was (before it was cut by surface 1-2-3-4-5-6-7-8) shaped like a backward C and it is important to realize that the object is still basically shaped like a backward C. (Note the left-side view.) The fact that the object contains an oblique surface that cuts through several other surfaces need not complicate the drawing of orthographic views. Look back at Section 7-3, which illustrated and explained how to draw oblique surfaces, and compare the solution to Figure 7-6 with the solution to Figure 7-9. With the exception of the horizontal slot in Figure 7-9, the problems are the same.

Figure 7-11 is another example of a problem that involves parallel edges.

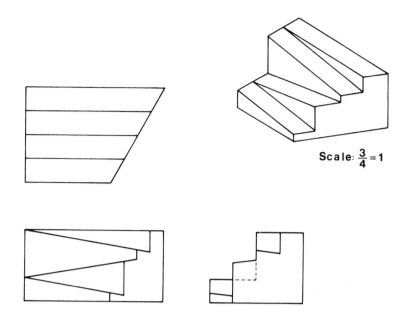

Scale: $\frac{3}{4} = 1$

FIGURE 7-11   Three views of an object that contains several sets of parallel edges.

7-5 DIHEDRAL ANGLES

Figure 7-12 is a problem that involves a dihedral angle. A dihedral angle is an angle between two planes. The problem is to draw the front view of the object when the top and right-side views are given. Figure 7-13 is the solution and was derived by the following procedure:

GIVEN: Top and right-side views.
PROBLEM: Draw the front view.

FIGURE 7-12

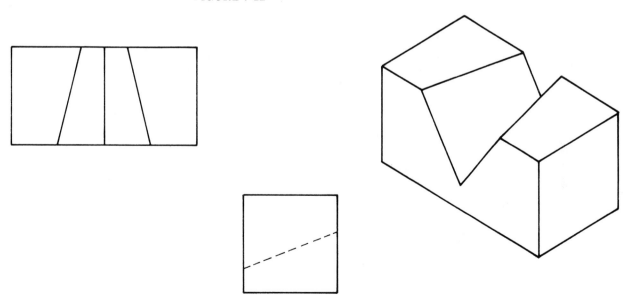

SOLUTION:

1. Make, to the best of your ability, a freehand sketch of the solution and, if possible, a sketch of the entire object.

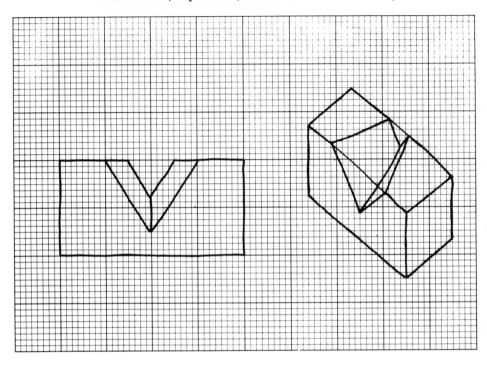

FIGURE 7-13(a)

2. Define the vortex line of the dihedral angle. In this example, the vortex line is defined as line 1-2.

FIGURE 7-13(b)

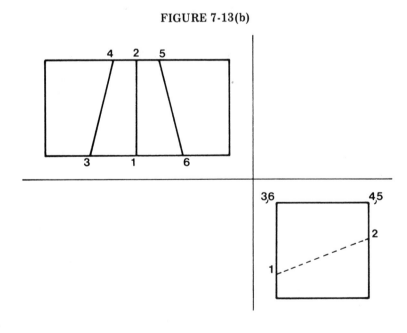

3. Define the surfaces that make up the dihedral angle. In this problem, the surfaces are 3-4-1-2 and 1-2-5-6.

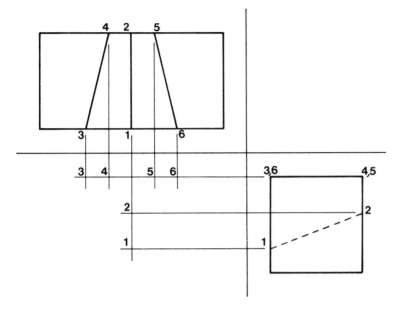

FIGURE 7-13(c)

4. Project points 1, 2, 3, 4, 5, and 6 into the front view by using projection theory.

FIGURE 7-13(d)

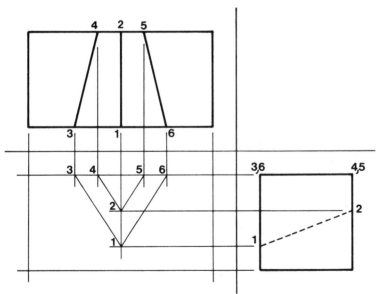

5. Using very light lines, lay out the front view of surfaces 3-4-1-2 and 1-2-5-6. After checking your work, complete the initial layout of the entire project.

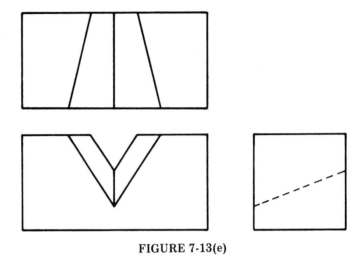

FIGURE 7-13(e)

6. Erase all excess lines and darken in all lines to their final color and configuration.

Figure 7-14 is another example of a problem that involves a dihedral angle.

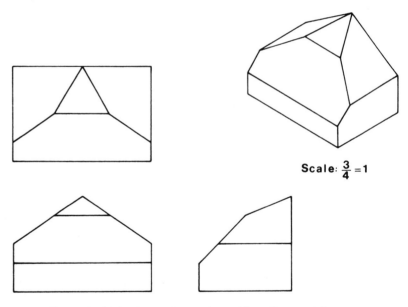

Scale: $\frac{3}{4} = 1$

FIGURE 7-14 Three views of an object that contains several dihedral angles.

7-6   HOLES IN OBLIQUE SURFACES

Figure 7-15 is a problem that involves a hole in an oblique surface. The problem is to draw the top view of the object when the front and right-side views are given. Figure 7-16 is the solution and was derived by the following procedure:

GIVEN: Front and right-side views.
PROBLEM: Draw the top view.

FIGURE 7-15

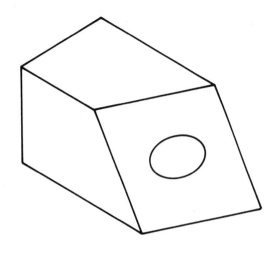

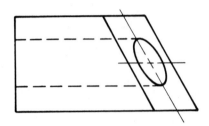 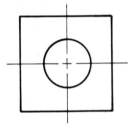

SOLUTION:

1. Make, to the best of your ability, a sketch of the solution and, if possible, the entire object.

FIGURE 7-16(a)

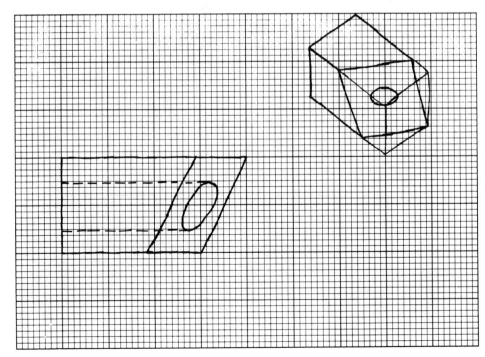

2. Using very light lines, draw the top view (not including the hole) by using the procedure outlined for oblique surfaces in Section 7-3.

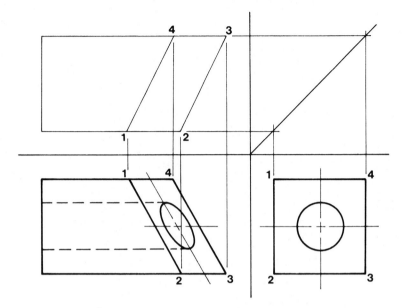

FIGURE 7-16(b)

3. In the right-side view, where the hole appears as a circle, mark off and label points 5 through 16 at 30° intervals around the circle. Although these points do not really exist on the circle, they are to be used for reference.

FIGURE 7-16(c)

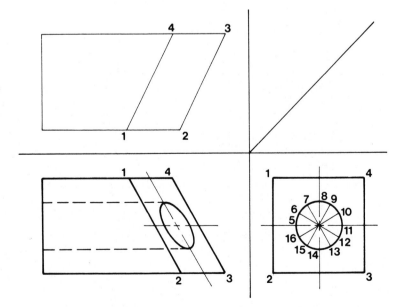

4. Project points 5 through 16 from the right-side view into the
    front view as shown in Figure 7-16(d). Label the points. Be
    careful not to reverse the points when you project between
    views. For example, points 16 and 12 are on the same horizontal
    projection line, but point 16 is to the left of center and point
    12 is to the right of center.

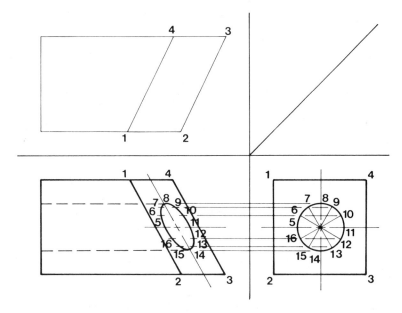

FIGURE 7-16(d)

5. Using the information from the front and right-side views, pro-
    ject points 5 through 16 into the top view. Check each point
    carefully.

FIGURE 7-16(e)

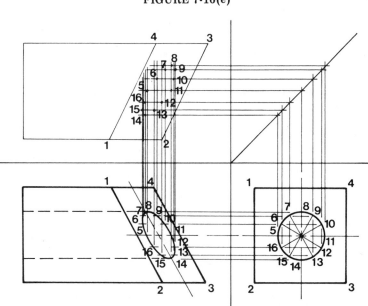

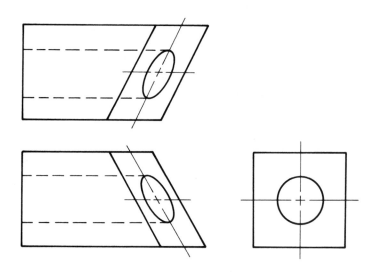

FIGURE 7-16(f)

6. Erase all excess lines and darken in all lines to their final color and configuration.

The use of $30°$ intervals in step 3 was made simply because it is easy to draw $30°$ angles with a T-square and a 30–60–90 triangle as a guide. Any angle could have been used, including randomly spaced angles. The more points used, the more accurate will be the projected ellipse.

But what if we must work from an isometric drawing? Figure 7-17 shows an isometric drawing of the object used for Figure 7-15, but this time we know less about the shape of the hole because we are given much less information to work with. Nevertheless, we can draw three views of the object, including the hole. Figure 7-18 is the drawing sequence used to convert the isometric drawing given in Figure 7-17 to three orthographic views. The following procedure was used:

GIVEN: An object.
PROBLEM: Draw front, top, and right-side views.

SOLUTION:

1. Draw the front, top, and right-side views of the object from the given information. You will not be able to include the hole.

FIGURE 7-18(a)

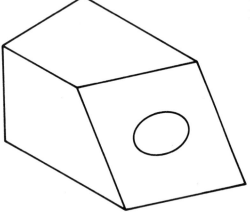

FIGURE 7-17

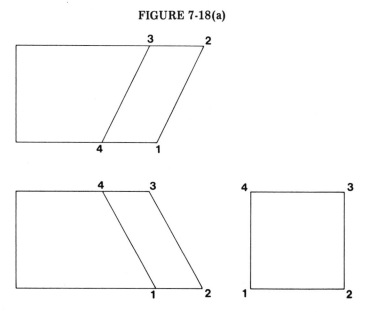

2. Since we know that the hole will be drilled in a horizontal direction and that it will be centered in the object, we can draw it as a circle in the right-side view.

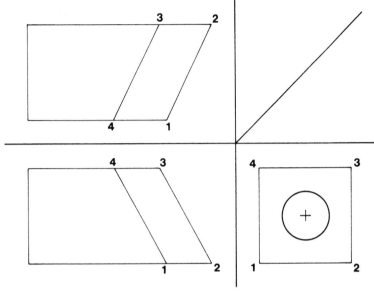

FIGURE 7-18(b)

3. Using a T-square and 30–60–90 triangle, mark off lines, 30° apart, in the right-side view as shown. Label the lines.

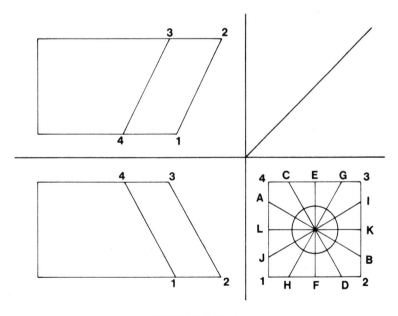

FIGURE 7-18(c)

4. Project lines A-I, L-K, and J-B into the front view by projecting points A, L, and J from line 4-1 in the side view to line 4-1 in the front view and points I, K, and B from line 3-2 in the side to line 3-2 in the front view. Note that lines A-I, L-K, and J-B cannot be projected into the top view and

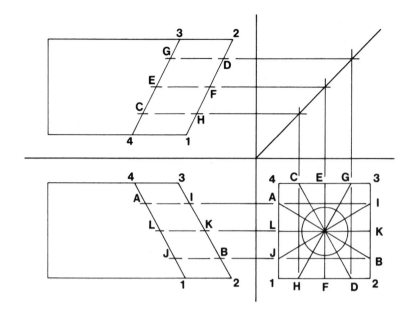

FIGURE 7-18(d)

that lines C-D, E-F, and G-H cannot be projected into the front view.

5. Project points A, L, J, I, K, and B from the front view to the top view as shown.

FIGURE 7-18(e)

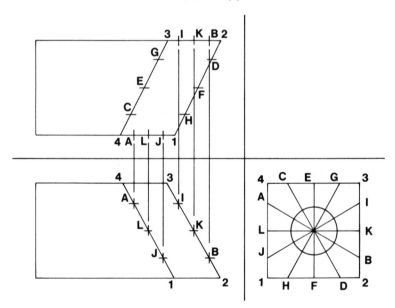

6. Project points C, E, G, H, F, and D from the top view into the front view as shown.

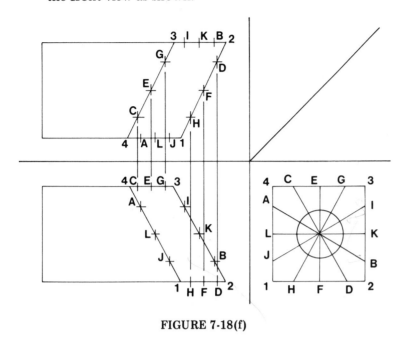

FIGURE 7-18(f)

7. Draw in lines A-B, C-D, E-F, G-H, I-J, and L-K in the front and top views.

FIGURE 7-18(g)

8. In the side view, label the intersections that the 30° lines make with the side view of the hole.

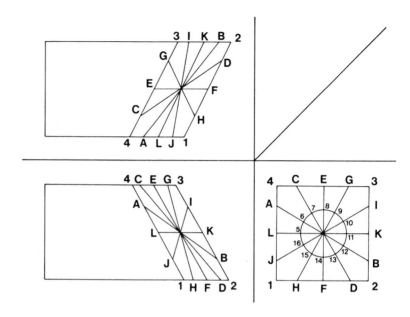

FIGURE 7-18(h)

9. Project points 6, 7, 8, 9, 10, 12, 13, 14, 15, and 16 into the front view and points 9, 10, 11, 12, 13, 15, 16, 5, 6, and 7 into the top view. Note that points 5 and 11 cannot be projected into the front view and that points 8 and 14 cannot be projected into the top view.

FIGURE 7-18(i)

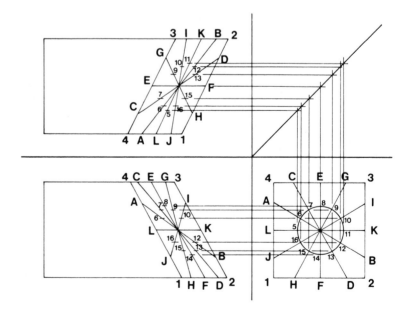

10. Project points 8 and 14 from the front view to the top view and project points 5 and 11 from the top view to the front view.

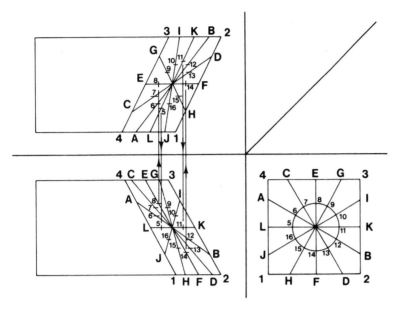

FIGURE 7-18(j)

11. The hole is now defined in each view. Erase all excess lines and darken in the final lines (including the hole) to their proper color and configuration.

FIGURE 7-18(k)

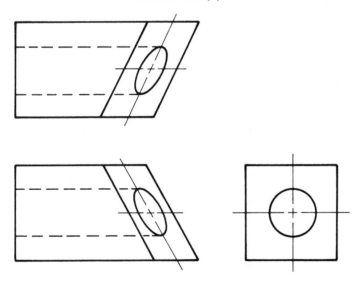

7-7 INTERNAL SURFACES
IN OBLIQUE SURFACES

Figure 7-19 is a problem that involves an internal surface in an oblique surface. This kind of problem is very similar to problems that involve holes in oblique surfaces. The problem here is to draw the front view of the object when the top and right-side views are given. Figure 7-20 is the solution and was derived by the listed procedure:

GIVEN: Top and right-side views.
PROBLEM: Draw the front view.

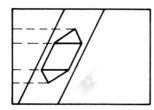

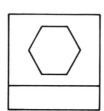

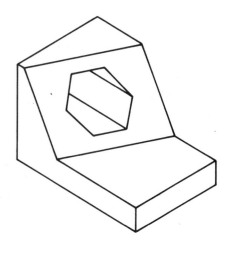

FIGURE 7-19

SOLUTION:

1. Draw a front view of the object. Include the oblique surface and omit the internal surfaces. Use the outline presented in Section 7-3.

FIGURE 7-20(a)

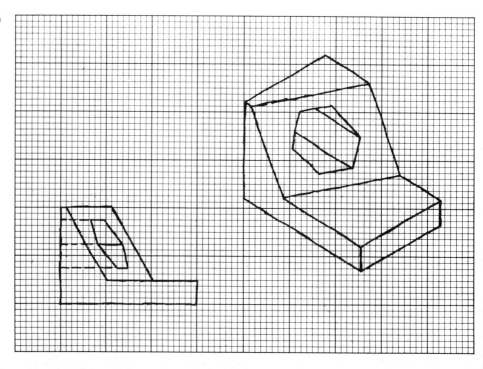

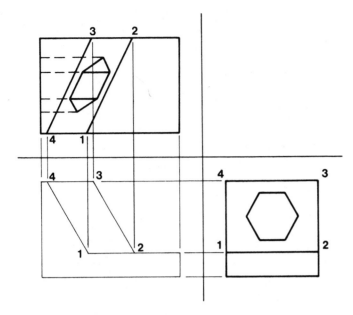

FIGURE 7-20(b)

2. Number the points of the internal surfaces in the given views. In this example, the six corners of the hexagon cutout were labeled points 5, 6, 7, 8, 9, and 10.

FIGURE 7-20(c)

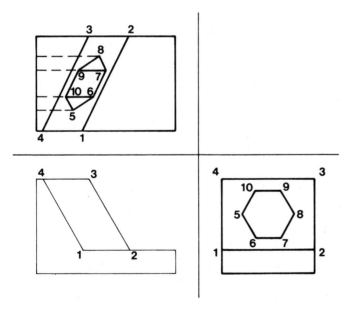

3. Using projection theory, project points 5 through 10 into the front view.

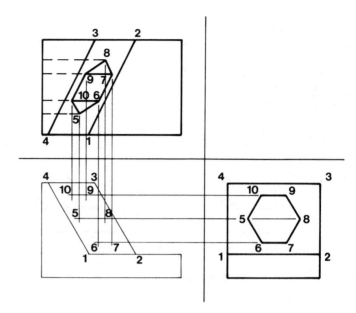

FIGURE 7-20(d)

4. Erase all excess lines and darken in all lines to their final color and configuration.

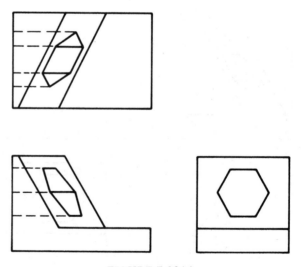

FIGURE 7-20(e)

When you work on a problem that involves internal surfaces, it is important that you carefully label the intersection of the internal surfaces with the outer surfaces. If necessary, add imaginary points (as was done for holes in internal surfaces) to help ensure an accurate projection of the shape of the intersection.

PROBLEMS

P7-1   Draw three views (front, top, and side) of each object.
through
P7-6

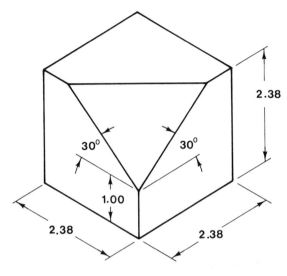

FIGURE P7-1

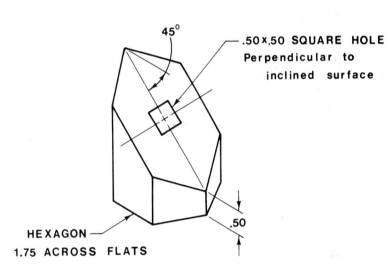

45°

.50×.50 **SQUARE HOLE**
Perpendicular to
inclined surface

.50

HEXAGON
1.75 ACROSS FLATS

FIGURE P7-2

FIGURE P7-3

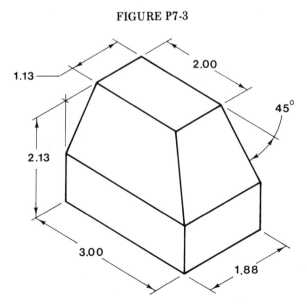

FIGURE P7-4

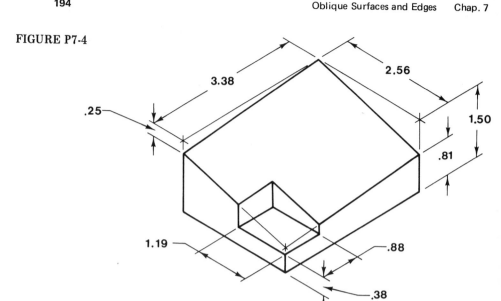

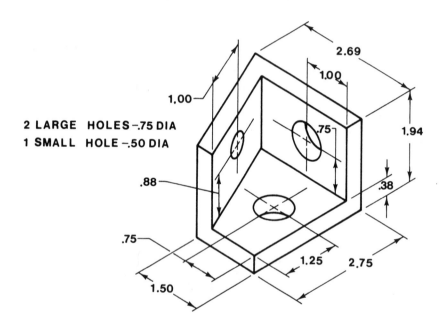

2 LARGE HOLES –.75 DIA
1 SMALL HOLE –.50 DIA

FIGURE P7-5

FIGURE P7-6   All dimensions are in millimeters.

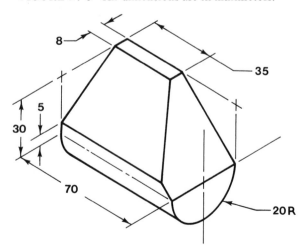

P7-7
through
P7-17
For each figure, redraw the given two views and add the appropriate view so that each object is defined by a front, top, and side view. Each square of the grid pattern is 0.20 on each side.

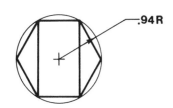

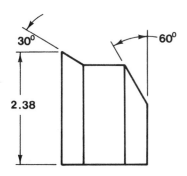

FIGURE P7-7

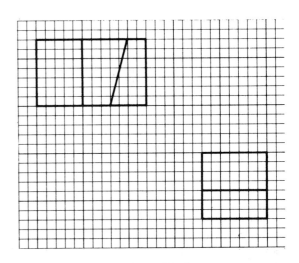

FIGURE P7-10

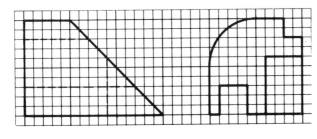

FIGURE P7-8

FIGURE P7-11

FIGURE P7-9

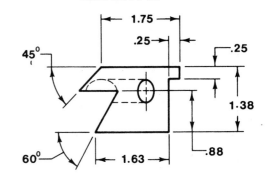

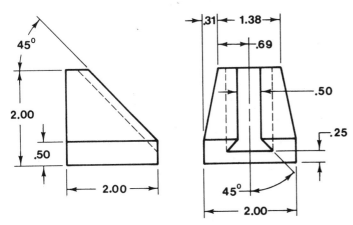

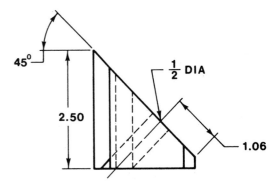

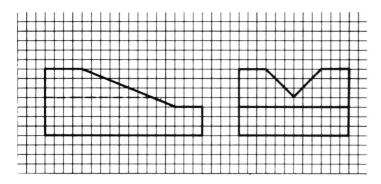

FIGURE P7-12

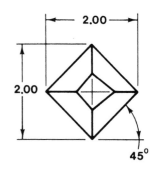

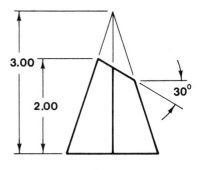

FIGURE P7-13

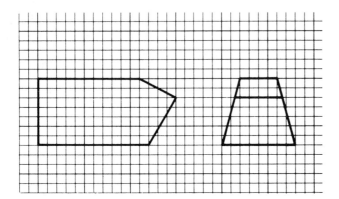

FIGURE P7-14　This problem is based on information supplied courtesy of TRW Carbide Division.

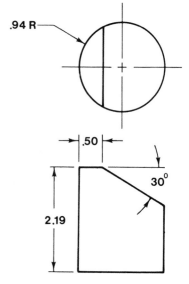

FIGURE P7-15

FIGURE P7-16

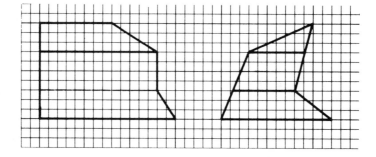

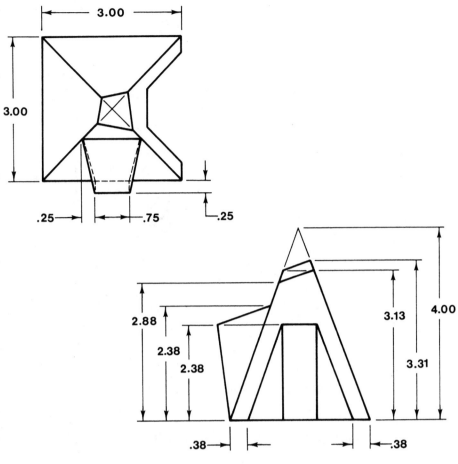

FIGURE P7-17   Problem courtesy of Tony Lazaris.

P7-18   Draw a front, top, and right-side view of the following objects.
through
P7-20

FIGURE P7-18

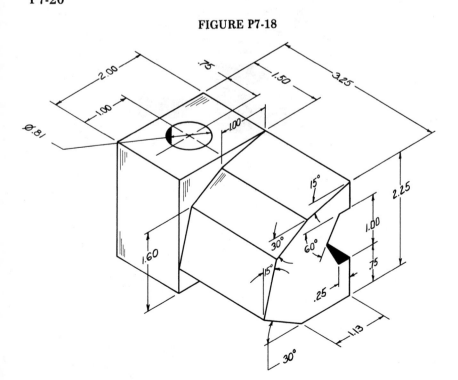

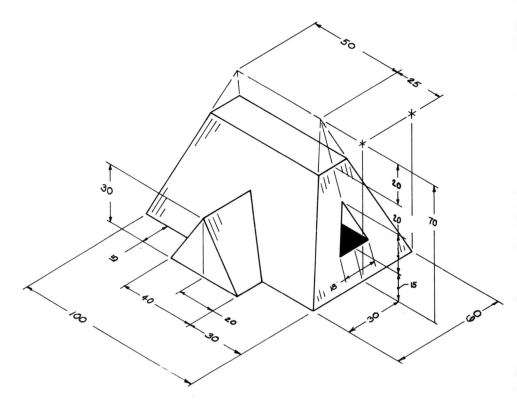

FIGURE P7-19   All dimensions are in millimeters.

FIGURE P7-20   All dimensions are in millimeters.

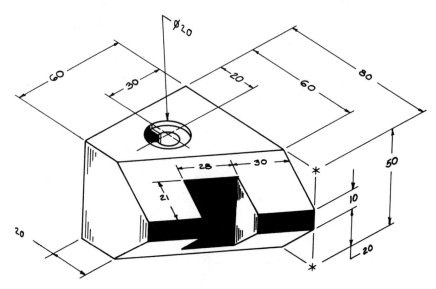

8

CYLINDERS

FIGURE 8-0

8-1   INTRODUCTION

Cylinder problems are problems whose basic geometric shape is a cylinder. They are often difficult to visualize and draw because they have no natural flat surfaces, making it confusing to know where to start. Figure 8-1 demonstrates this point by showing front, top, and right-side views of a natural, uncut cylinder. The front and top views are identical and regardless of how the cylinder is rotated about the center point, x, the front and top views remain identical. How can we label or reference cylinders to make sure that those who read the finished drawings clearly understand which view is the front and which is the top?

The key to solving the problems is in using the centerlines. The right-side view of Figure 8-1 defines vertical and horizontal centerlines which divide the cylinder into four equal quadrants. Where the horizontal and vertical centerlines cross the periphery of the cylinder are defined as *centerline edge points*, and are marked points 1, 2, 3, 4, 5, 6, 7, and 8. They are all double points and represent the end views of longitudinal centerlines which can be seen in the front and top views. These longitudinal centerlines 1-2, 3-4, 5-6, and 7-8 can be used to define the cylinder's height and width and can be used as theoretical baselines from which to reference variances from the basic cylindrical shape (cuts, chamfers, and so on).

It should be understood that although longitudinal centerlines do not physically exist on cylindrical pieces, they represent where the curved surface of the cylinder changes direction (see Section 5-5).

Centerlines will be used throughout this chapter, as they are in industry, to define and give reference to cylinder problems. The first step in any cylinder problem should be to define the centerlines.

FIGURE 8-1   Three views of a cylinder.

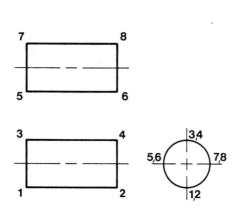

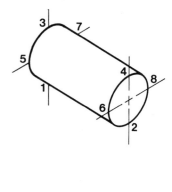

8-2   CUTS ABOVE THE CENTERLINE

Figure 8-2 is an example of a cylinder cut lengthwise above the centerline. The problem is to find the top view when the front and right-side
views are given. Figure 8-3 is the solution and was derived by the following procedure:

GIVEN: Front and right-side views.
PROBLEM: Draw the top view.

FIGURE 8-2   Cylinder cut above the centerline.

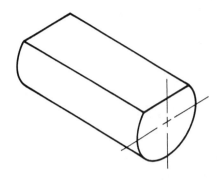

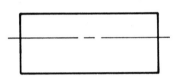

SOLUTION:

1. Define the horizontal centerline edge points—5, 6, 7, and 8—
    in the front and right-side views.

FIGURE 8-3(a)

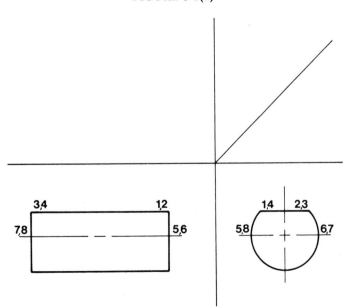

2. Project points 5, 6, 7, and 8 into the top view; then, using construction lines, connect the points to form a rectangle.

3. Define the cut surface 1, 2, 3, and 4 in the front and right-side views.

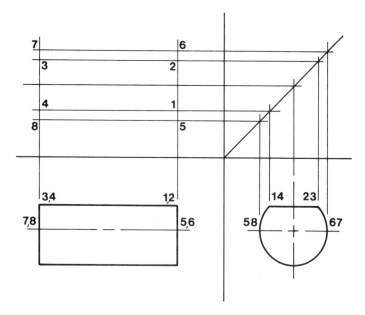

FIGURE 8-3(b)

4. Project points 1, 2, 3, and 4 into the top view and, using construction lines, connect the points to form a rectangle.

5. Darken in the two rectangles with object lines.

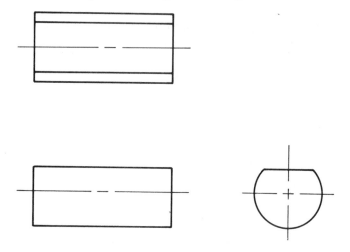

FIGURE 8-3(c)

Although a technical solution to the problem has been derived, there still may be some difficulty in visualizing what it means. The cut surface 1-2-3-4 which appears as a rectangle in the top view, appears as a straight line in both the front and right-side views. The surfaces 8-5-1-4 and 3-2-6-7 appear in the top view to be similar to the flat surface 1-2-3-4, but they are not. Surfaces 8-5-1-4 and 3-2-6-7 are curved surfaces that start at horizontal centerlines 5-8 and 7-6 and extend upward to lines 4-1 and 7-6. Study the right-side view to verify the length, height, and shape of the curve. Remember that although center edge lines 8-5 and 7-6 do not really appear on the piece, they represent the widest part of the cylinder and where the curve defining the periphery changes directions from outward to inward (see Section 4-7).

8-3   CUTS BELOW THE CENTERLINE

Figure 8-4 is an example of a cylinder cut lengthwise below the centerline. The problem is to find the top view when the front side views are given. Figure 8-5 is the solution and was derived by the following procedure:

GIVEN: Front and side views.
PROBLEM: Draw the top view.

FIGURE 8-4   Cylinders cut below the centerline.

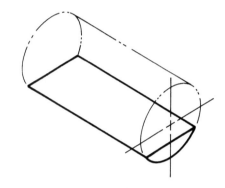

SOLUTION:

1. Define the four corners of the cut surface 1-2-3-4 in the front and top views.

FIGURE 8-5(a)

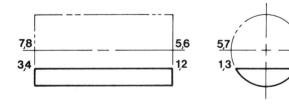

2. Project points 1, 2, 3, and 4 into the top view.

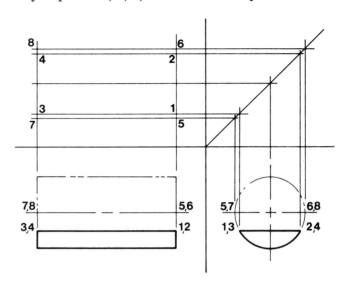

FIGURE 8-5(b)

3. Connect points 1, 2, 3, and 4 with object lines to complete the top view.

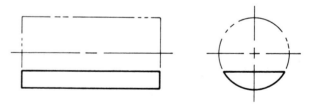

FIGURE 8-5(c)

The plane 1-2-3-4 is a flat surface that has a round surface directly under it.

8-4   INCLINED CUTS

Figure 8-6 is an example of an inclined cut. The problem is to find the top view when the front and right-side views are given. Figure 8-7 is the solution and was derived by the following procedure:

GIVEN: Front and right-side views.
PROBLEM: Draw the top view.

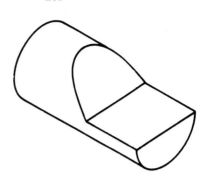

FIGURE 8-6 Cylinder with an inclined cut.

SOLUTION:

1. Define the horizontal centerline edge points 1, 9, 10, and 11 in the front and right-side views.

2. Project the horizontal centerline edge points 1, 9, 10, and 11 into the top view, thereby defining the outside edge of the cylinder.

3. Create points 2, 3, 4, 5, 6, 7, and 8 in the right-side view by marking off angles of 0°, 30°, 60°, 90°, 60°, 30°, and 0° from the horizontal centerline (30° increments were chosen because they are easy to draw with a 30-60-90 triangle). These points are for drawing purposes only and do not represent any corners

FIGURE 8-7(a)

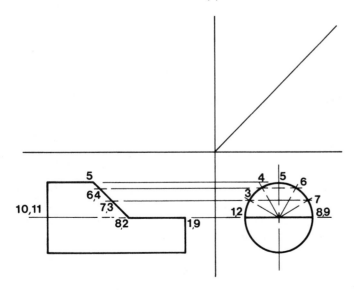

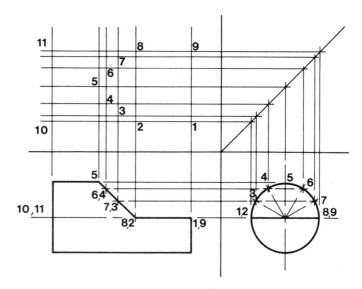

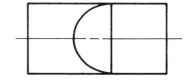

FIGURE 8-7(b)

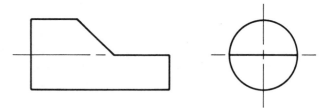

FIGURE 8-7(c)

or edges which appear on the piece and, therefore, they should be drawn very lightly. Once the solution has been derived, these points should be erased.

4. Project points 2, 3, 4, 5, 6, 7, and 8 into the front view. This is done by drawing lines parallel to the horizontal principal plane line from the created points 2, 3, 4, 5, 6, 7, and 8 to the inclined surface in the front view. Points (6, 4), (7, 3), and (8, 2) become double points in the front view.

5. Project points 2, 3, 4, 5, 6, 7, and 8 into the top view.

6. Using a French curve, carefully draw in the elliptical shape by connecting points 2, 3, 4, 5, 6, 7, and 8. Be careful to avoid a lumpy or ragged curve. The finished ellipse should be smooth and symmetrical.

7. Complete the top view by projecting the necessary points from the front and right-side views.

8-5 CURVED CUTS

Figure 8-8 is an example of a cylinder with a curved cut. The problem is to find the top view when the front and right-side views are given. Figure 8-9 is the solution and was derived by the listed procedure:

GIVEN: Front and right-side views.
PROBLEM: Draw the top view.

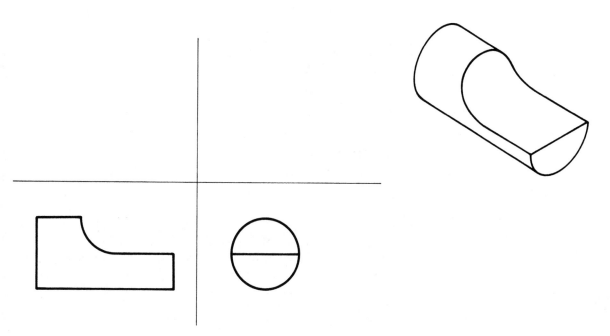

FIGURE 8-8 Cylinder with a curved cut.

SOLUTION:

1. Define the horizontal centerline edge points 1, 2, 3, and 4 in the front and right-side views.

FIGURE 8-9(a)

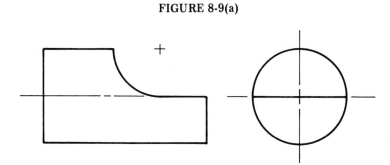

2. Project the horizontal centerline edge points 1, 2, 3, and 4 into the top view, thereby defining the outside edge of the cylinder.

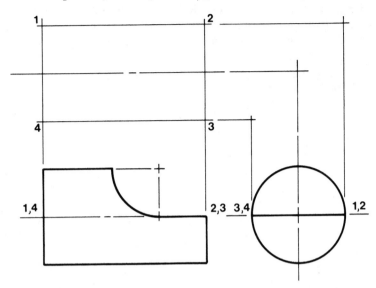

FIGURE 8-9(b)

3. Create points 5, 6, 7, 8, 9, 10, and 11 in the right-side view by marking off angles of $0°$, $30°$, $60°$, $90°$, $60°$, $30°$, and $0°$ from the horizontal centerline ($30°$ increments were chosen because they are easy to draw with a 30–60–90 triangle). These points are for drawing purposes only and would never appear on the piece; therefore, they should be drawn very lightly. After the solution has been derived, these points should be erased.

FIGURE 8-9(c)

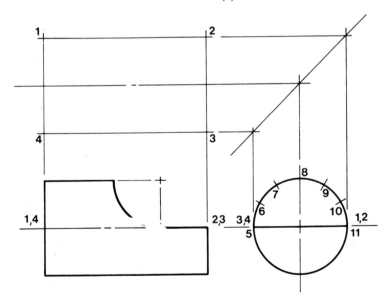

4. Project points 5, 6, 7, 8, 9, 10, and 11 into the front view. This
    is done by drawing lines parallel to the horizontal principal
    plane line from the created points 5, 6, 7, 8, 9, 10, and 11 to
    the inclined surface in the front view. Points (5, 11), (6, 10),
    and (7, 9) become double points in the front view.

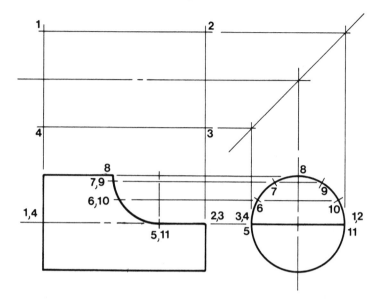

FIGURE 8-9(d)

5. Project points 5, 6, 7, 8, 9, 10, and 11 into the top view.

FIGURE 8-9(e)

6. Using a French curve, carefully draw in the elliptical shape by connecting points 5, 6, 7, 8, 9, 10, and 11. Be careful to avoid a lumpy or ragged curve. The finished ellipse should be smooth and symmetrical.

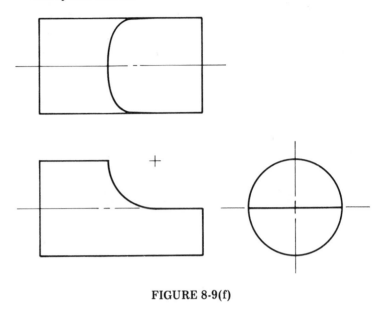

FIGURE 8-9(f)

7. Complete the top view by projecting necessary points from the front and right-side views.

This procedure is exactly the same as that used in Section 8-4 to solve inclined-cut problems. However, because of the difference in the kind of cut, the resulting ellipse in the top view is different.

8-6   CHAMFERS

Chamfers are machine cuts, usually at 45°, along the edges or corners of machined pieces. They are used to eliminate sharp, dangerous edges, to trim off material for clearance requirements, or to act as a kind of taper in aligning parts. They are not unique to cylinder problems and Figure 8-10 gives two examples of chamfers in noncylindrical pieces. Figure 8-11 is an example of cylindrical chamfers. The problem is to find the top view where the front and right-side views are given. Only the chamfered sections are labeled since the rest of the solution was explained previously (Section 8-1).

Consider line 2-3-4-5 in the right-side view of Figure 8-12. It is an end view of a flat plane that was developed by machining away part of the cylinder and then chamfering the end. The chamfer creates two edge lines that show as concentric circles in the right side view and as parallel lines in the front view. This means that points 3 and 4 are in front of points 2 and 5 and that lines 2-3 and 4-5 are slanted in the top view. After defining points 2, 3, 4, and 5 in the front and side views, project them into the top view and draw lines 2-3, 3-4, and 4-5. There is no line 2-5. Why?

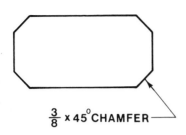

$\frac{3}{8}$ x 45° CHAMFER

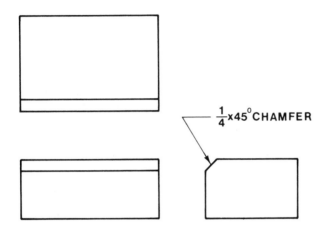

$\frac{1}{4}$ x 45° CHAMFER

FIGURE 8-10 Chamfers on noncylindrical shaped objects.

GIVEN: Front and right-side views.
PROBLEM: Draw the top view.

FIGURE 8-11 Cylinder with chamfers.

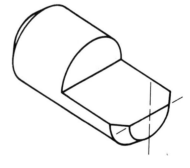

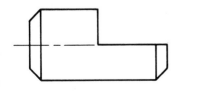

SOLUTION:

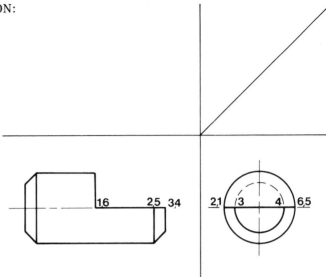

FIGURE 8-12(a)

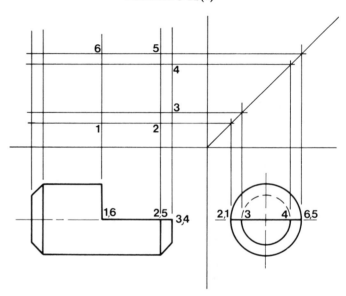

FIGURE 8-12(b)

FIGURE 8-12(c)

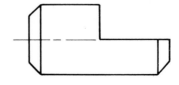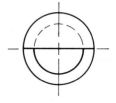

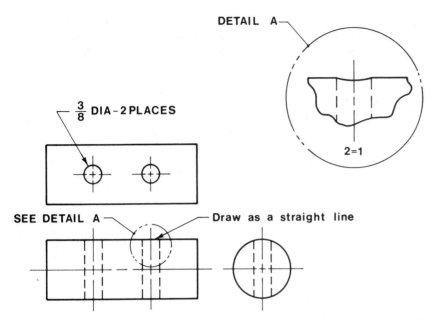

FIGURE 8-13   How to draw holes in cylinders.

8-7   HOLES

When holes are drilled in cylinders, holes create unique drawing and projection problems. Figure 8-13 is an example of a cylinder that has two holes drilled completely through from top to bottom. Detail A is an enlargement of the top surface and has been drawn twice scale to accent the elliptical shape generated by the round hole. Even at twice scale, the ellipse is almost flat. Thus, in most drawings the ellipse is neglected and is drawn as a straight line as shown in Figure 8-13. This irregularity is acceptable drafting practice since it does not affect the accuracy of the communication. A machinist need only to know the size and location of the hole. The fact that a slight elliptical shape is generated in an orthographic projection will not affect the drilling procedure. The ellipse is an unimportant result of the drilling and may be omitted.

This is not true if the hole is large. Where is the crossover point? When does a hole become large enough to require an ellipse to be drawn? There is no fixed rule to follow and drafters must use their own discretion depending on their particular situation.

8-8   ECCENTRIC CYLINDERS

Eccentric cylinders are two or more cylinders whose center points are not matched. One cylinder is off center in relation to the other. Some students feel that eccentric problems are created by instructors who are eccentric, but no research has been done to prove or, for that matter, disprove this theory.

Eccentric problems should be approached as separate and independent cylinder problems. Break down the problems into the sections that make them up and solve them separately; then rejoin them to form a composite solution. Figure 8-14 is an example of an eccentric cylinder problem that requires a top view when front and right-side views are given. Figure 8-15 is the solution and was derived by the listed procedure:

GIVEN: Front and right-side views.
PROBLEM: Draw the top view.

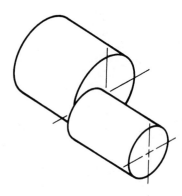

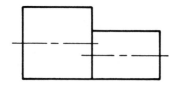 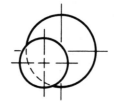

FIGURE 8-14   Eccentric cylinders.

SOLUTION:

1. Define the centerline edge points 1, 2, 3, and 4 of the smaller diameter cylinder in the front and right-side views.
2. Project points 1, 2, 3, and 4 into the top view.

FIGURE 8-15(a)

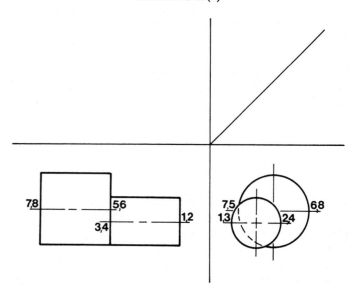

3. Define the centerline edge points 5, 6, 7, and 8 of the larger diameter cylinder in the front and right-side views.

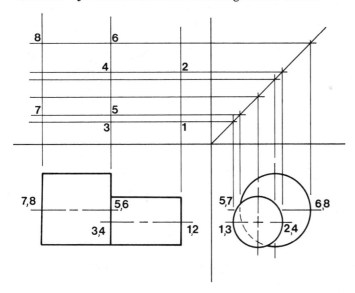

FIGURE 8-15(b)

4. Project points 5, 6, 7, and 8 into the top view.

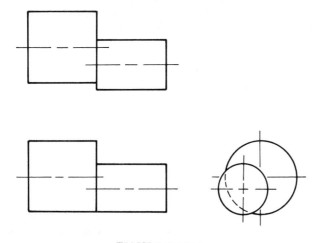

FIGURE 8-15(c)

5. Draw in the appropriate visible lines.

8-9 HOLLOW SECTIONS

Figure 8-16 is an example of a hollow cylinder. The problem is to find the top view when the front and right-side views are given. Figure 8-17 is the solution and was arrived at by considering the outside and inside diameters as separate cylinders, solving them independently, and forming a composite solution. The following steps were used:

GIVEN: Front and right-side views.
PROBLEM: Draw the top view.

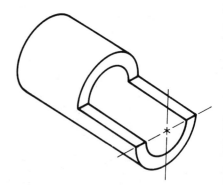

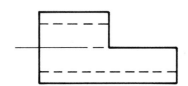 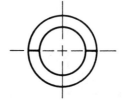

FIGURE 8-16 Hollow cylinders.

SOLUTION:

1. On the outside cylinder, define the horizontal centerline edge
    points of the cut surface 1-2-7-8 in the front and right-side
    views. In other words, consider the problem to consist only of
    a solid cylinder, cut directly on the horizontal centerline (see
    Section 8-2).

FIGURE 8-17(a)

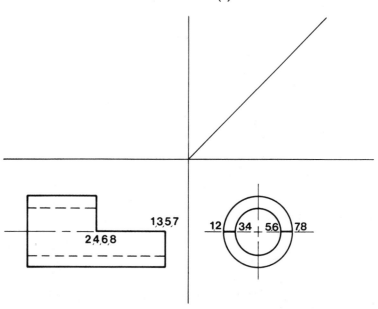

2. Repeat step 1 for the inside cylinder, defining points 3, 4, 5, and 6.

3. Project points 1, 2, 3, 4, 5, 6, 7, and 8 into the top view.

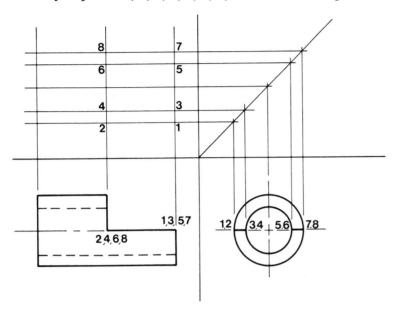

FIGURE 8-17(b)

4. Draw in the surfaces 1-2-3-4 and 5-6-7-8. Note that surfaces 1-2-6-5 and 3-4-6-7 are flat rectangles and, with the exception of the cylinder's ends, are the only flat surfaces in the problems.

5. Complete the top view by projecting necessary points from the front and right-side views.

FIGURE 8-17(c)

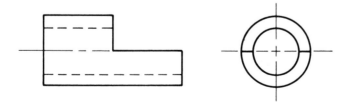

PROBLEMS

P8-1   Draw three views (front, top, and right side) of each object.
through If two views are given, redraw the given views and add the
P8-19   missing third view. Make a freehand three-dimensional sketch
of the object if requested by your instructor. If a three-
dimensional picture is used to present the object, draw the
front, top, and right-side views. Each square on the grid pattern is 0.20 per side.

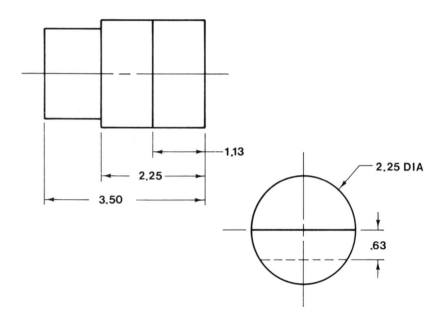

FIGURE P8-1

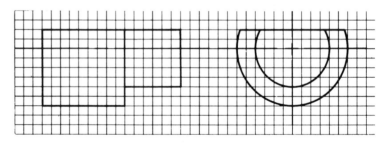

FIGURE P8-2

FIGURE P8-3

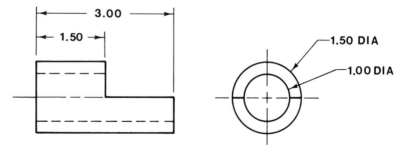

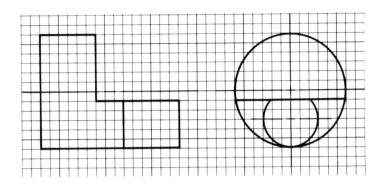

FIGURE P8-4

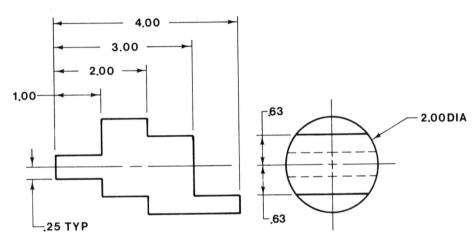

FIGURE P8-5

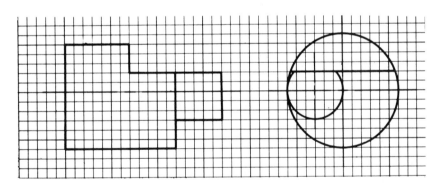

FIGURE P8-6

FIGURE P8-7

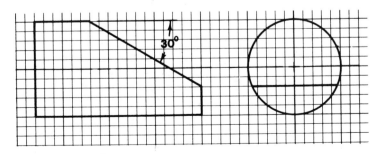

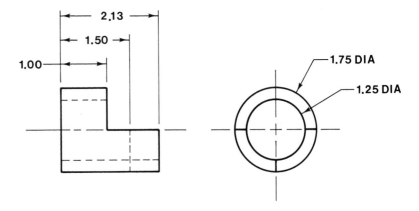

FIGURE P8-8

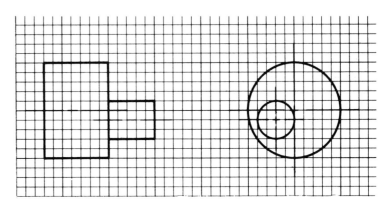

FIGURE P8-9

FIGURE P8-10

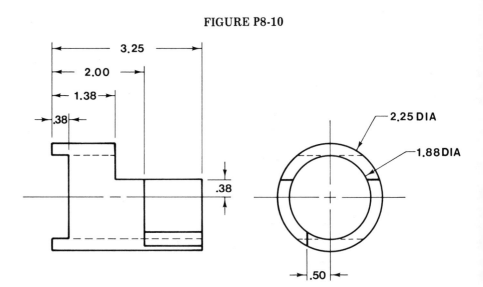

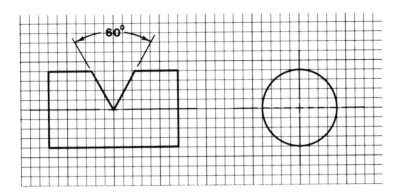

FIGURE P8-11

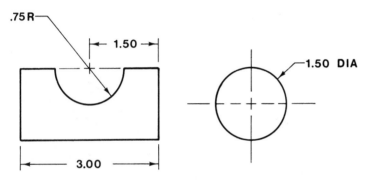

FIGURE P8-12

FIGURE P8-13   All dimensions are in millimeters.

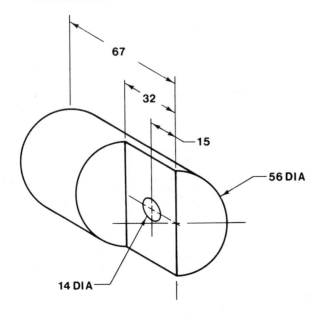

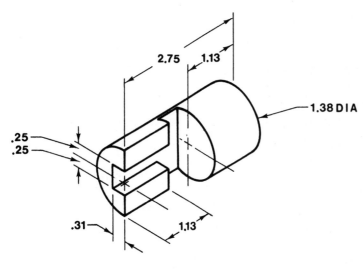

FIGURE P8-14

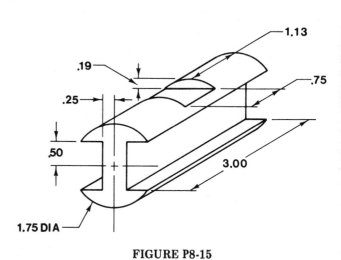

FIGURE P8-15

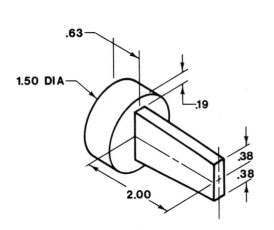

FIGURE P8-16

FIGURE P8-17   Use B-size paper (11 × 17).

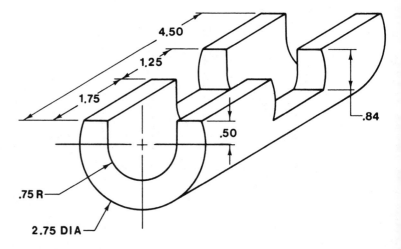

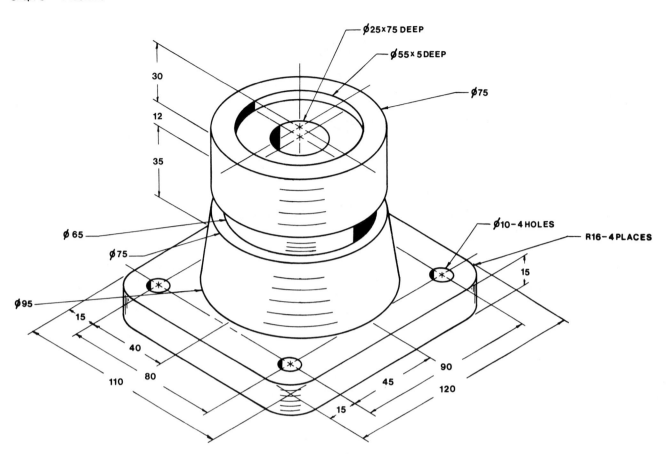

FIGURE P8-18   All dimensions are in millimeters.

FIGURE P8-19   All dimensions are in millimeters.

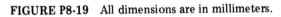

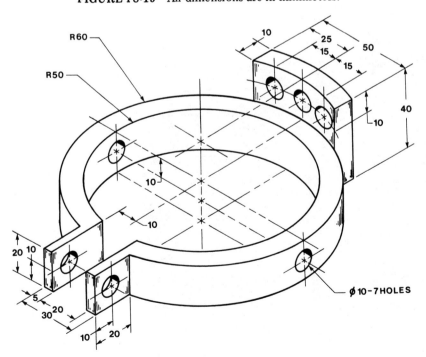

9

CASTINGS

FIGURE 9-0

9-1 INTRODUCTION

Objects that are made using the casting process present unique drawing problems. Because edges of cast objects are not square (90°), they
cannot appear as lines in orthographic views. Also, these nonsquare
edges often intersect each other, which results in many unusually shaped
lines. This chapter presents the techniques used to draw cast objects
and shows how rounded edges may be represented.

9-2 FILLETS AND ROUNDS

A fillet is a concave-shaped edge. A round is a convex-shaped edge.
Figure 9-1 illustrates these definitions. The size of a fillet or round is
usually specified on a drawing by a note such as

$$\text{ALL FILLETS AND ROUNDS } \tfrac{1}{8}\text{R}$$

although they may be dimensioned individually.

From a drawing standpoint, fillets and rounds only appear in views
taken at 90° to them, as shown in Figure 9-2. Note in Figure 9-2 that
the lines that *seem* to represent edges represent surfaces. The actual
edges are rounded and so do not appear in orthographic views unless
they are shown in profile.

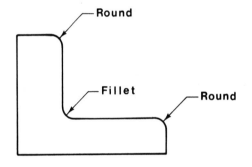

FIGURE 9-1 Fillets and rounds.

FIGURE 9-2 Orthographic views of a cast object.

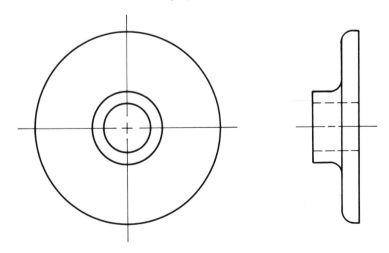

ALL FILLETS & ROUNDS $\tfrac{1}{4}$R

225

Most fillets and rounds are drawn by using a circle template because they are too small to be easily drawn with a compass. Remember that when you use a circle template the hole sizes are given in diameters, not radii. Therefore, be sure to convert the given fillet and round sizes to diameters before you draw them.

9-3   ROUNDED-EDGE REPRESENTATION

When you draw cast objects, how do you properly represent rounded edges? Should the small curved lines, as shown in Figure 9-3(a) be used? Should the long, phantom lines, as shown in Figure 9-3(b), be used? Or should no lines be used, as shown in Figure 9-3(c)?

In general, the small curved lines and the phantom lines are only used to indicate a rounded edge in a pictorial drawing. They are not used in orthographic views. However, since representation practices vary from company to company, always check the company standards before you start a drawing.

FIGURE 9-3   No special shading or line work is required to represent the rounded edges of a cast object.

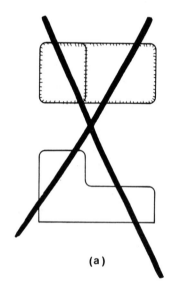

(a)

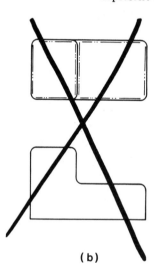

(b)

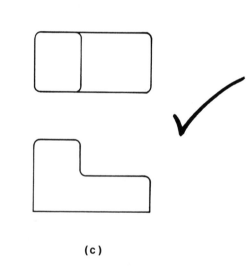

(c)

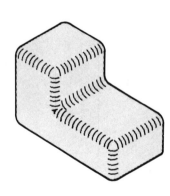

9-4 RUNOUTS

A runout is the intersection of two or more rounded edges. Runouts appear on a drawing as curved sections at the end of the lines that represent surfaces. They generally turn out (that is, away from the surface lines), but this is not a hard-and-fast rule. Elliptical surfaces generate runouts that turn in, as illustrated in Figure 9-4. Each object must be judged individually as to which runout direction looks the most realistic.

Figure 9-5 shows several different examples of runouts. Draw runouts either freehand or by using a curve as a guide.

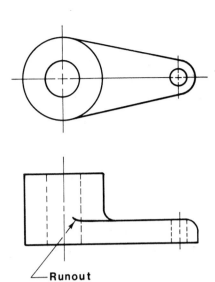

FIGURE 9-4 Drawing that includes a runout.

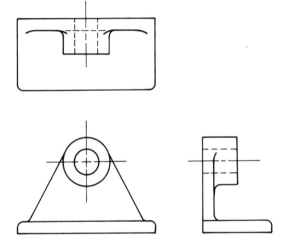

FIGURE 9-5 Example of two objects whose orthographic views include runouts.

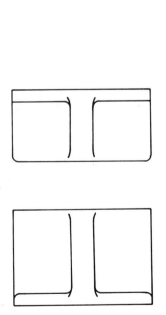

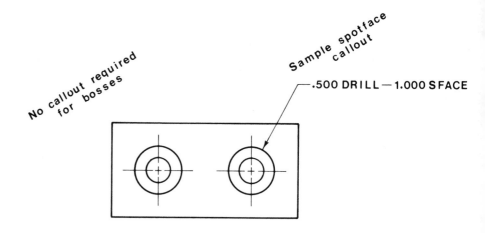

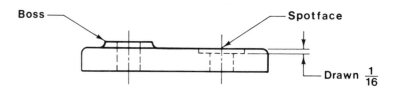

FIGURE 9-6   Drawing that includes a spotface and a boss.

9-5   SPOTFACES AND BOSSES

Spotfacing is a special machining operation that smoothes out the otherwise rough surface finish found on cast objects. It is similar to counterboring, but during spotfacing the surface of the object is cut just deep enough to produce a machined quality finish.

Spotfacing is called out on a drawing by a note as shown in Figure 9-6. First, the drill diameter is given, then the drill depth, if any, and finally the diameter of the spotface. Spotface depth is *not* specified unless it is a design requirement. The machinist will cut just deep enough into the object to smooth out the surface.

When you draw a spotface, draw the spotface depth 1/16 inch. This depth enables the drawing reader to clearly see the spotface and is convenient to draw. Other parts of the note are interpreted as shown in Figure 9-6.

A boss is a raised portion of a casting, as shown in Figure 9-6. Bosses are usually added to castings because they can be easily machined (being higher than the rest of the cast surface). Bosses are usually as high as the given fillet and round size.

9-6   MACHINING MARKS

Machining marks are used to differentiate those surfaces on a casting which are to be machined. Figure 9-7 illustrates different machining marks and shows how they are used.

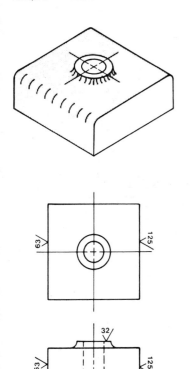

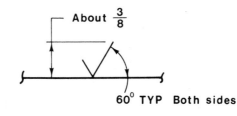

$500 \vee$ = Rough

$250 \vee$ = Coarse

$125 \vee$ = Medium

$63 \vee$ = Fine

$32 \vee$ = Very fine

About $\frac{3}{8}$

60^0 TYP Both sides

FIGURE 9-7 Machining marks are used to indicate the quality of the surface finish required.

PROBLEMS

P9-1   Draw three views (front, top, and side) of each object.
through
P9-3

FIGURE P9-1

All fillets and rounds $= \frac{3}{16}$ R

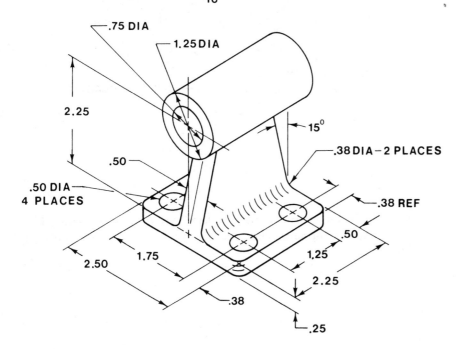

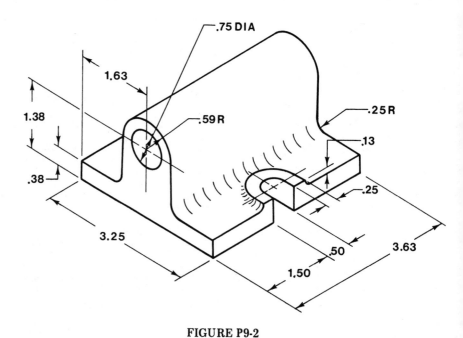

FIGURE P9-2

All fillets and rounds = $\frac{1}{8}$ R

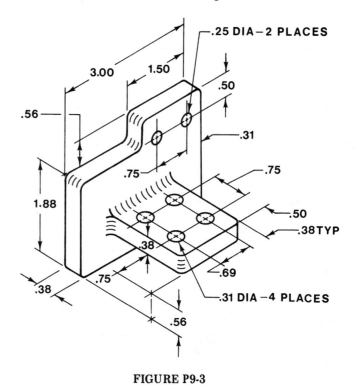

FIGURE P9-3

P9-4
through
P9-9
For each problem, redraw the two views given and add the required missing view. Each square on the grid pattern is 0.20 per side.

TOP

All fillets and rounds = $\frac{3}{16}$R

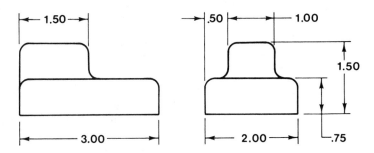

FIGURE P9-4

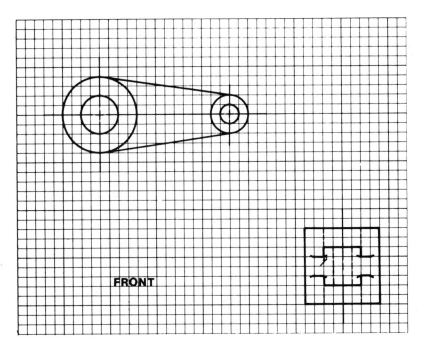

FIGURE P9-5

FIGURE P9-6

TOP

All fillets and rounds = $\frac{1}{8}$R

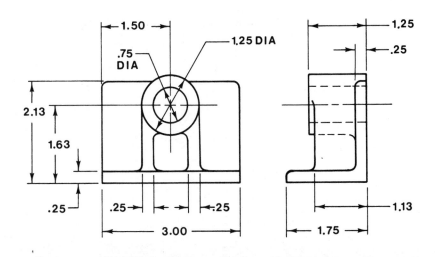

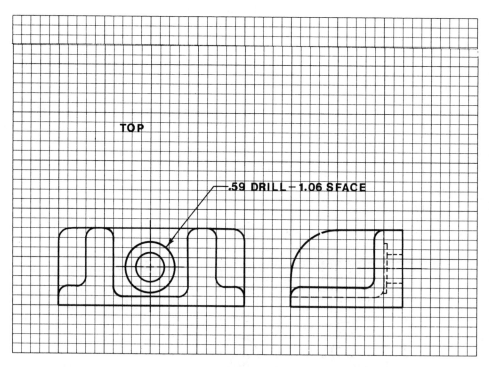

TOP

.59 DRILL – 1.06 SFACE

FIGURE P9-7

FIGURE P9-8

TOP

All fillets and rounds $= \frac{1}{8}$ R

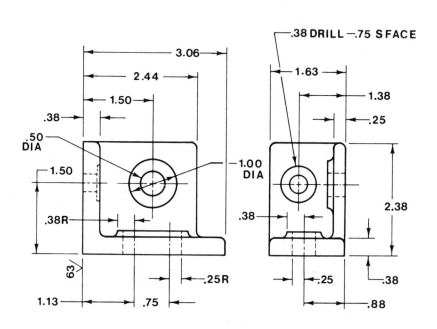

.38 DRILL –.75 SFACE

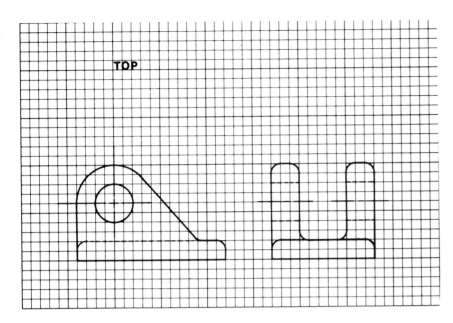

FIGURE P9-9

P9-10
through
P9-14

Draw a front, top, and side view of the following objects.

FIGURE P9-10   All dimensions are in millimeters.

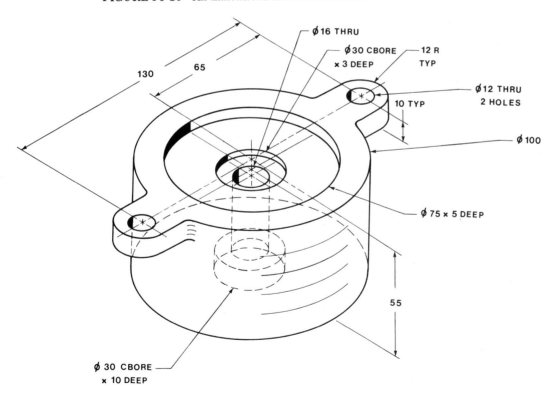

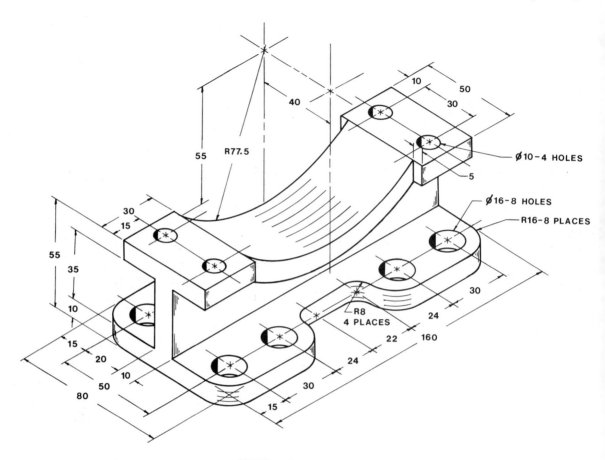

FIGURE P9-11 All dimensions are in millimeters.

FIGURE P9-12 All dimensions are in millimeters.

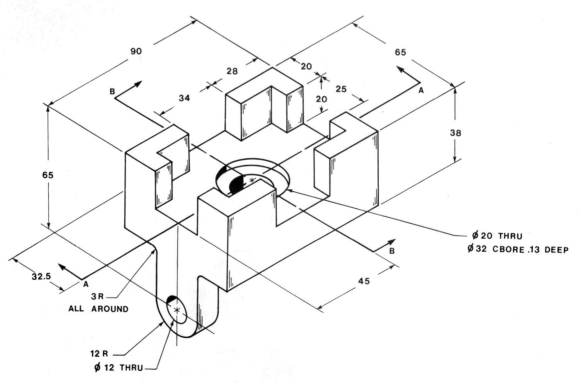

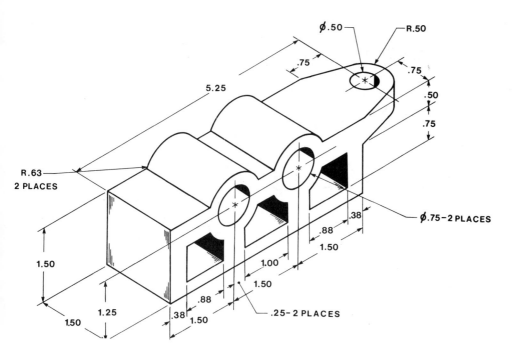

FIGURE P9-13   All dimensions are in inches.

FIGURE P9-14   All dimensions are in inches.

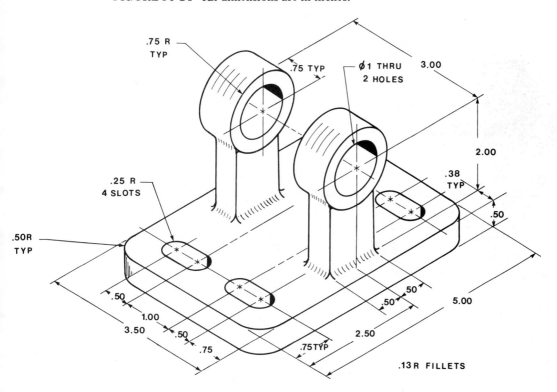

10

SECTIONAL VIEWS

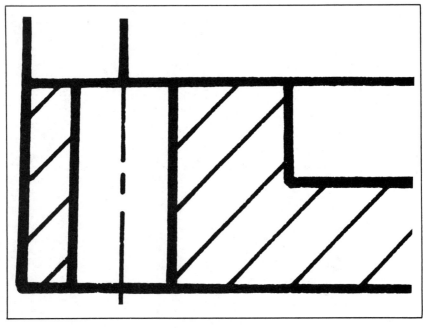

FIGURE 10-0

10-1   INTRODUCTION

Sectional views are used to expose internal surfaces of an object that would otherwise be hidden from direct view. Sectional views greatly add to the clarity of a drawing because they do not contain any hidden lines.

To help you understand the differences between section cuts and regular orthographic views, study Figures 10-1 and 10-2. In Figure 10-1 note the clarity of the internal profile of the object shown in the sectional view. In Figure 10-2 note that the sectional views are much easier to understand than are the regular orthographic views which contain many hidden lines. This does not mean that sectional views should be used instead of regular orthographic views. Sometimes sectional views may be used to replace confusing regular orthographic views. At other

FIGURE 10-1   Comparison between a regular orthographic view and a sectional view.

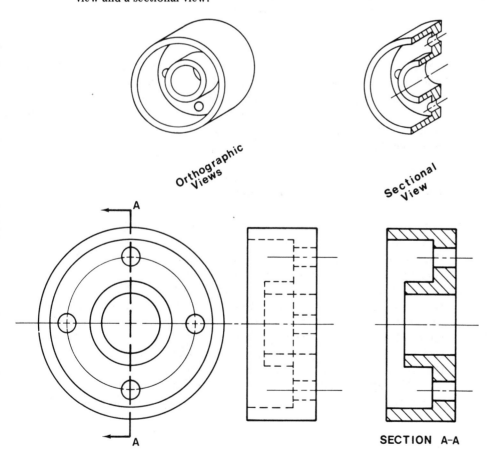

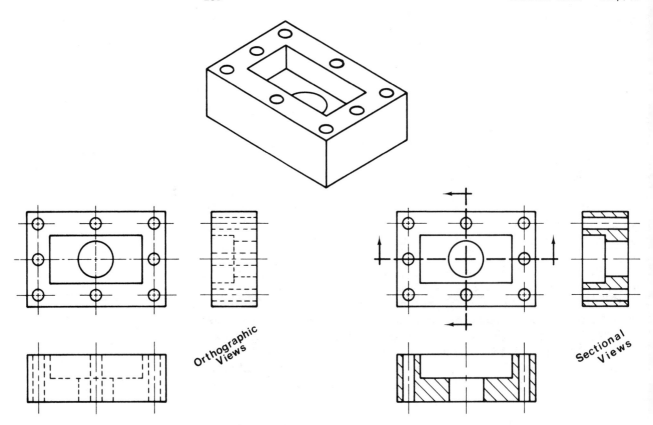

FIGURE 10-2 Comparison between regular orthographic views and sectional views.

times both views may be used. There are also many drawings that would not require section cuts at all. When and where to use a sectional view depends on the object being drawn. However, in any situation, your prime concern should be that your drawings are easily understood. Always be as clear and direct as possible in the views that you present.

10-2   CUTTING PLANE LINES

Cutting plane lines are used to define the line along which an object is to be cut. They are drawn by using either of the two configurations shown in Figure 10-3. They should be drawn by using very heavy and very black lines — as heavy and black, if not more so, than visible lines.

Cutting plane lines need not go directly through an object but may be offset as shown in Figures 10-4 and 10-5. Cutting plane lines are offset so that several internal surfaces may be shown in the same sectional view. The fact that a cutting plane line is offset does not appear in the sectional view. There should be no lines in the sectional view to indicate that the cutting plane line has changed direction.

The arrowheads of a cutting plane line indicate the direction in which to observe the sectional view. The actual section view should be

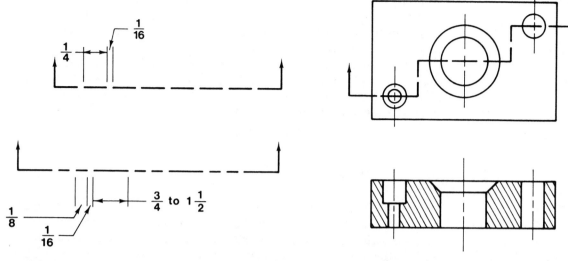

FIGURE 10-3 Cutting plane line configurations.

FIGURE 10-4 Offset cutting plane line.

Angular Cutting Plane Line

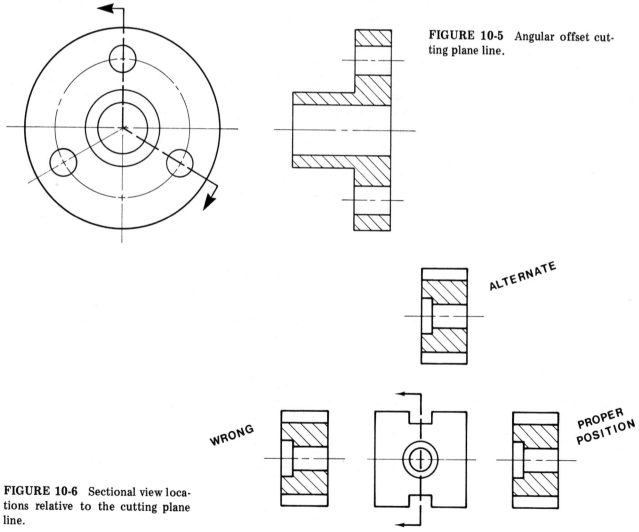

FIGURE 10-5 Angular offset cutting plane line.

FIGURE 10-6 Sectional view locations relative to the cutting plane line.

located behind the arrowheads or, if absolutely necessary, in alignment with the cutting plane line as illustrated in Figure 10-6. Under no circumstances should the sectional view be placed ahead of the cutting plane line arrowheads.

To help you visualize this convention, think of yourself as standing on the sectional view looking at the object being drawn. The cutting plane line arrowheads should point in the direction in which you are looking — away from the sectional view.

10-3   SECTION LINES

Section lines are used to indicate where, in a sectional view, solid material has been cut. There are many different section line patterns (a different pattern for each building material), but the most common pattern is the one shown in Figure 10-7. The lines are thin and black (about one-half as thick as visible lines) and are drawn at any inclined angle ($45°$ is the most often used).

When two or more parts are cut by the same cutting plane line, the section cut lines must be varied to indicate clearly the different parts. Section lines may be drawn at different angles or with different spacing, as illustrated in Figure 10-8.

Do not draw section lines so that they are parallel to any surface in the object. For example, the upper right corner of the object pictured in Figure 10-9 is a $45°$ surface. It is wrong to draw section lines at $45°$, parallel to the $45°$ surface. The lines must be drawn at another angle so that they are not parallel to the $45°$ surface.

There are several techniques drafters use to draw section lines. One is to use an Ames Lettering Guide. Another is to slip a piece of graph paper under the drawing, align it as desired, and then trace the lines. Another technique is to scribe a line onto a 45–45–90 triangle as shown in Figure 10-10. Scribe the guide line 1/8 inch from and parallel to the edge of the triangle. If desired, several lines may be scribed.

FIGURE 10-7   Section lines.

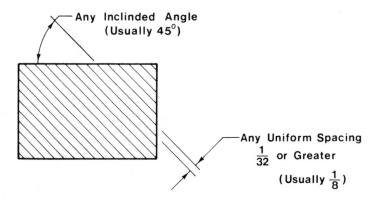

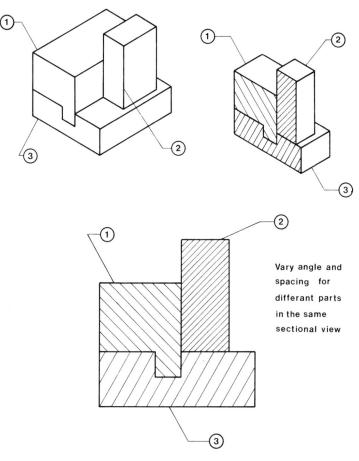

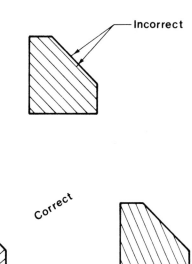

Vary angle and
spacing   for
differant parts
in the same
sectional view

FIGURE 10-8 A sectional view that cuts through several
objects.

FIGURE 10-9 Correct alignment of section lines.

FIGURE 10-10 A 45-45-90 triangle which has a line
scribed along the longest edge (see arrow). This scribed line
is parallel to and 1/8 from the edge of the triangle. Use the
scribed line to align the triangle when drawing section lines.

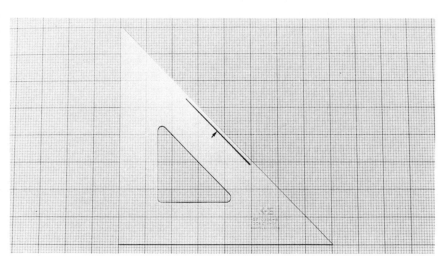

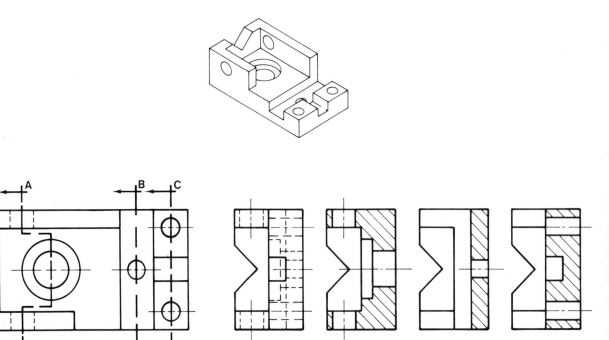

FIGURE 10-11   Multiple sectional views.

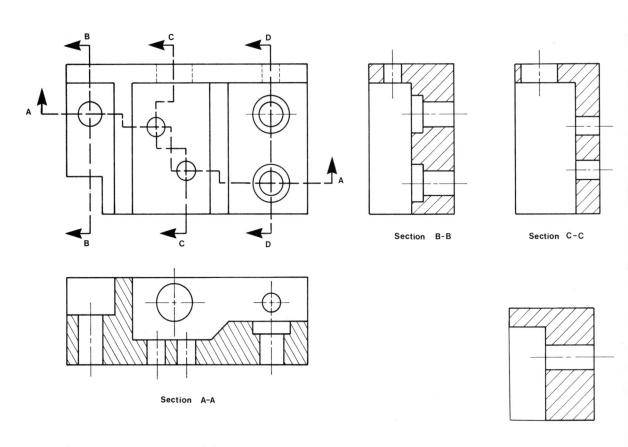

FIGURE 10-12   Multiple sectional views.

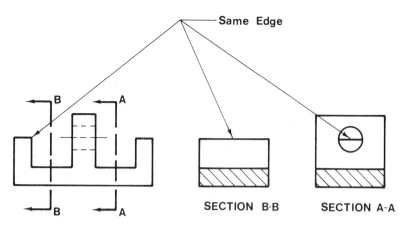

FIGURE 10-13   Multiple sectional views.

10-4   MULTIPLE SECTIONAL VIEWS

It is possible to take many sectional views through the same ortho-graphic view. Figures 10-11 and 10-12 demonstrate this by showing three sectional views, each taken through a different position of the same top view. Note how each sectional view is placed behind the arrowheads of the cutting plane line. As many sectional views as are necessary for clear definition of the object being studied may be shown.

Although hidden lines are not shown in sectional views, visible lines are shown. Note, for example, that the V formation located on the back left surface of the object appears in all three sectional views. Any surface that may be directly seen, even if it is not located directly on the path of the cutting plane line, must be shown. For example, the tall center portion of the object shown in Figure 10-13 appears in section A-A because it may be directly seen. The shorter end section cannot be directly seen and, therefore, is not shown. However, note that part of the shorter left end section may be directly seen through the hole and, because it may be seen, it must be shown in the sectional view. Since in section B-B we are beyond the tall center section, it will be omitted in the sectional view.

10-5   REVOLVED SECTIONAL VIEWS

It is sometimes possible to save drawing a separate sectional view by drawing a sectional view directly on the regular orthographic view. This sectional view is called a *revolved sectional view* and is illustrated in Figure 10-14. A revolved sectional view is used to define the shape of an object that has a constant shape.

FIGURE 10-14   Revolved sectional view.

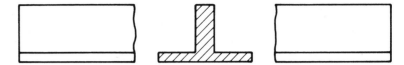

FIGURE 10-15 Revolved sectional view.

Figure 10-15 illustrates another revolved sectional view. This time the object has been broken open and the revolved sectional view has been placed between break lines. Either revolved sectional view (Figure 10-14 or Figure 10-15) is acceptable.

10-6 HALF SECTIONAL VIEWS

Regular orthographic views and sectional views may be combined within the same orthographic view to form a half sectional view. Figure 10-16(a) shows a half sectional view. Note that the two views are separated by a centerline and that each half is drawn independently of the other. The regular orthographic part of the view shows hidden lines, but the sectional view part does not. Half sections are particularly useful for drawing symmetrical objects.

Study the cutting plane line in Figure 10-16(a) and note how the left arrowhead is placed directly on the center line. Compare this with the cutting plane line of Figure 10-16(b) and then compare the differences in the resultant sectional views. By drawing the cutting plane line as shown in Figure 10-16(b), we elminate the need to draw all hidden lines on the left side of the sectional view. Figure 10-16(a) and (b) show two acceptable ways of drawing half sectional views.

FIGURE 10-16 Half sectional views.

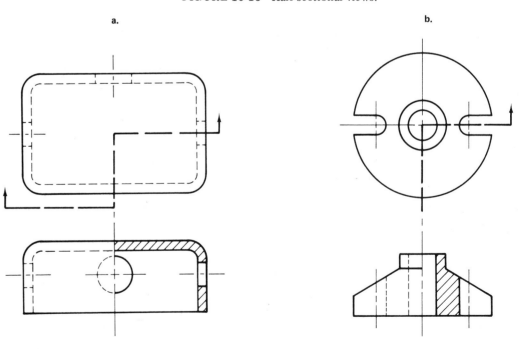

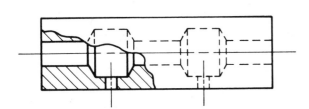

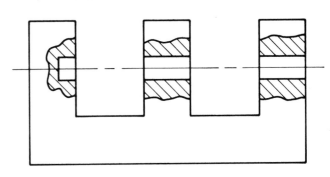

FIGURE 10-17 Broken-out sectional view.   FIGURE 10-18 Broken-out sectional view.

10-7   BROKEN-OUT SECTIONAL VIEWS

Sometimes less than a full or half sectional view is sufficient to clarify some internal surfaces of an object. In Figure 10-17, for example, the internal surfaces are symmetrical both vertically and horizontally. That is, the left and right halves are exactly the same as the top and bottom halves. Therefore, in this example we need only show a small piece of the internal surfaces to give the reader a good idea of the entire internal shape of the object. We do this by using a broken-out sectional view as shown in Figure 10-17.

Broken-out sectional views are sectional views drawn on a regular orthographic view and are created by theoretically breaking off a part of the external surface of the object, thereby exposing some of the internal surfaces to direct view. When you break open the object, use a break line to outline the place where the external surfaces have been broken. A cutting plane line is not required.

Figure 10-18 is another example of a broken-out sectional view.

10-8   PROJECTION THEORY

The projection theory presented in Chapter 4 and continued throughout this book is also applicable to sectional views. Figure 10-19 illustrates its application.

Most sectional views are drawn without the aid of projection theory, but as with regular orthographic views, projection theory is very helpful in checking lines.

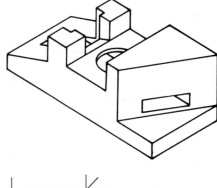

FIGURE 10-19 Multiple-sectional view problem solved using projection theory.

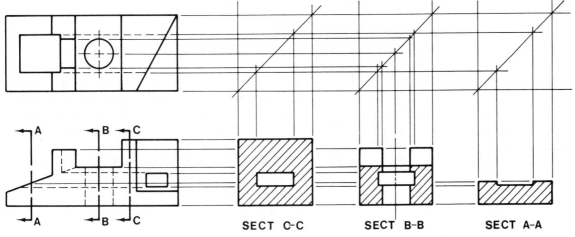

SECT C-C   SECT B-B   SECT A-A

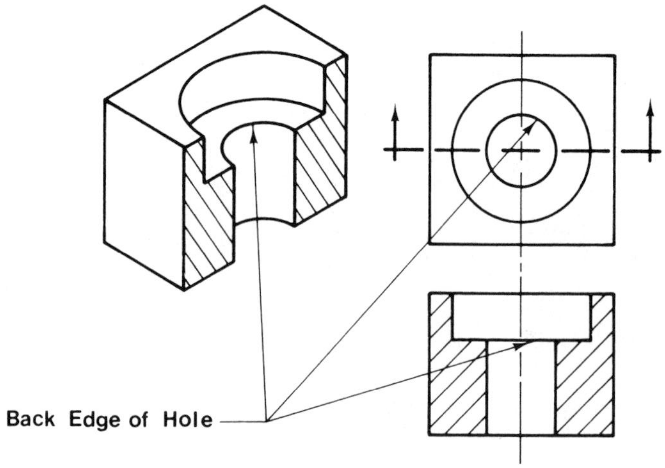

FIGURE 10-20   Hole in a sectional view.

10-9   HOLES IN SECTIONAL VIEWS

A common mistake that is made in drawing holes in a sectional view is to omit the back edge of the hole. Even if a hole is cut in half in a sectional view, the back edges must be shown in the sectional view. Figure 10-20, which shows a counterbored hole, uses an isometric drawing, a regular orthographic view, and a sectional view. In each view the arrows point to approximately the same point on the back edge of the hole. Note how lines that represent the back edges of the hole appear in the sectional view. When you draw holes in a sectional view, make sure that the back edge of the hole is represented.

10-10   AUXILIARY SECTIONAL VIEWS

Auxiliary sectional views may be created in the same way that auxiliary views are created (see Chapter 11). Use the cutting plane line to define the angle at which the view is to be taken and be sure to include sectioning lines where material has been cut. Either complete or partial auxiliary sectional views may be drawn. Figures 11-8 and 11-9 illustrate two auxiliary sectional views.

10-11   DIMENSIONING SECTIONAL VIEWS

Sectional views are very helpful in presenting clear, well-defined dimensions. In Chapter 6 we learned that it is considered poor practice to dimension to hidden lines. Yet there are many objects that contain so many internal surfaces that it is impossible to dimension without referring to hidden lines. By drawing sectional views, we open up to direct view the internal surfaces, thereby changing hidden lines to solid lines which, in turn, give us solid, well-defined lines to which to dimension.

Figure 10-21 illustrates how a sectional view may be dimensioned.

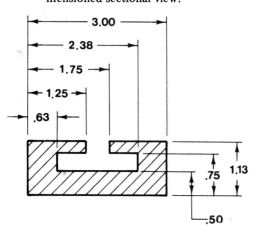

FIGURE 10-21   Example of a dimensioned sectional view.

Note that the extension lines cross over the section lines. Also note the small gap between the end of the extension line and the line that it is defining on the object.

10-12   SECTIONAL VIEWS OF ROUND OBJECTS

Figure 10-22 shows how sectional views are drawn for round solid and round tubular objects. In each example, an S-shaped break is used. An S-shaped break may be drawn using a template or by using the procedure outlined in Figure 10-23.

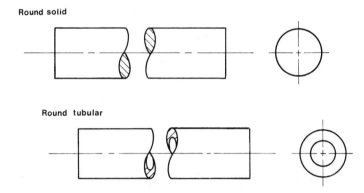

FIGURE 10-22   How to draw sectional views for round solid and tubular objects.

FIGURE 10-23   How to draw on S-shaped break.

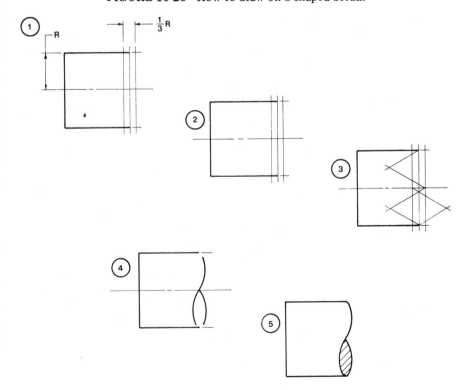

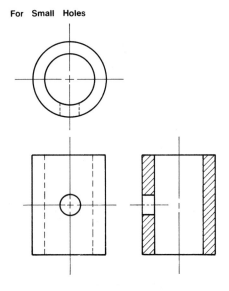

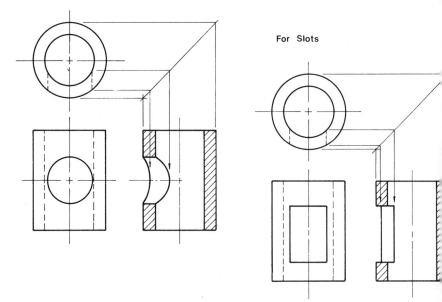

FIGURE 10-24 How to draw holes in tubular shaped objects shown in section.

An S-shaped break is drawn for any size of round object by first locating the center of the break. A distance equal to one-third of the object's radius is marked off on both sides of the break center as shown. Lines parallel to the break centerline are drawn. Thirty-degree lines are drawn as shown in step 3 of Figure 10-23 and their intersections are used as compass center points to draw part, but not all, of the S shape. The small arcs needed to complete the S shape from step 4 to step 5 are drawn freehand or by using an irregular curve as a guide.

10-13   HOLES IN TUBULAR OBJECTS

Sectional views in tubular-shaped objects that also include holes are drawn as shown in Figure 10-24. Large holes are drawn showing the correct ellipitical projection as shown. Small holes are drawn using straight lines. Slots are drawn using projected distances.

PROBLEMS

P10-1   Assume that each figure is a sectional view of a different
and P10-2   object. Redraw the sectional view and add section lines.
Each square on the grid is 0.20 per side.

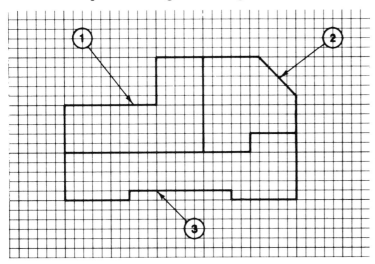

FIGURE P10-1

FIGURE P10-2

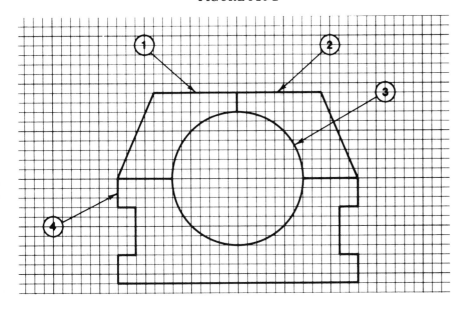

P10-3   Redraw the front view and replace the side view with a
through   sectional view. Each square on the grid is 0.20 per side.
P10-7

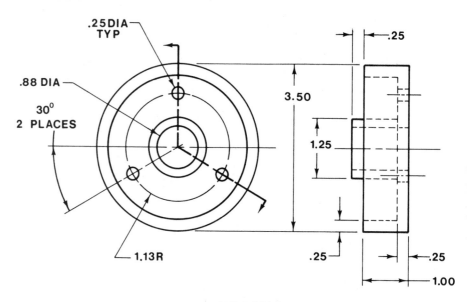

FIGURE P10-3

FIGURE P10-4

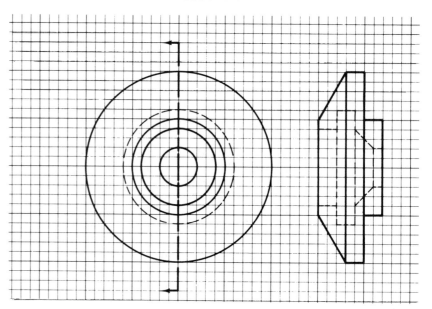

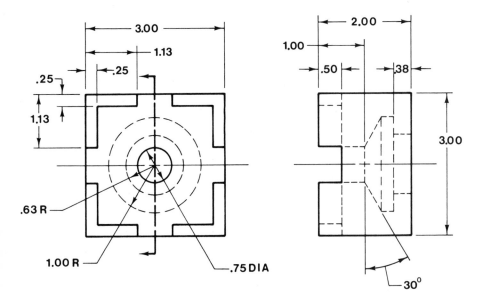

FIGURE P10-5

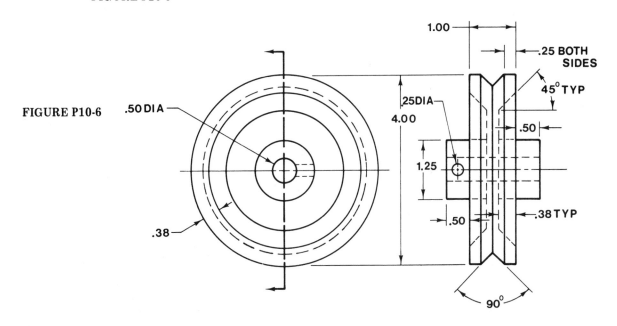

FIGURE P10-6

FIGURE P10-7

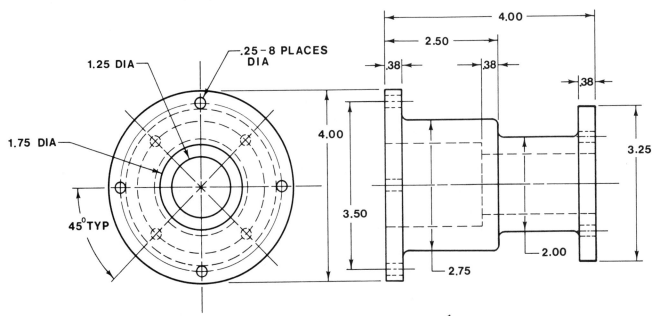

ALL FILLET AND ROUNDS = $\frac{1}{8}$R

P10-8   Redraw the top view shown in Figure P10-8 and replace the front view with a sectional view.

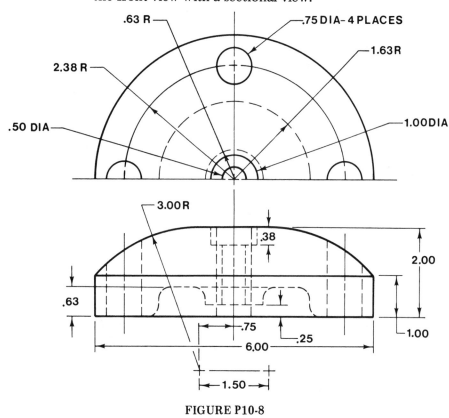

FIGURE P10-8

P10-9   Redraw the front and side views shown in Figure P10-9 and add the appropriate sectional views.

FIGURE P10-9

HOLE	x	y	DIA
A	1.50	2.00	.88
B	1.00	1.00	.63
C	2.50	1.00 / 2.00	.50
D	3.63	2.13	.75
E	3.25	1.13	.75

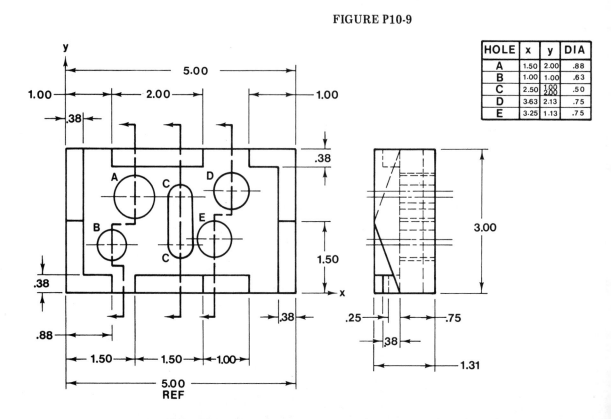

P10-10   Redraw the sectional view shown in Figure P10-10 and
         add the appropriate sectional views. Assume that all the
         pieces are round. Each square on the grid is 0.20 per side.

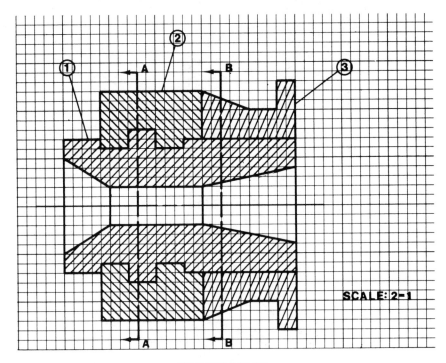

FIGURE P10-10

P10-11   Draw a front view and sectional view of each object. Each
through   triangle on the grid pattern is 0.20 per side.
P10-14

FIGURE P10-11

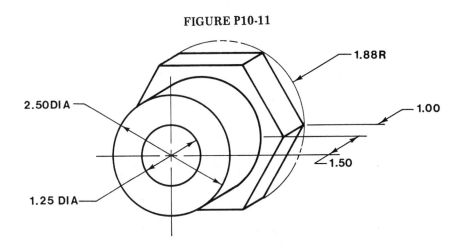

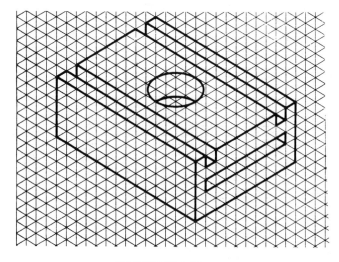

FIGURE P10-12

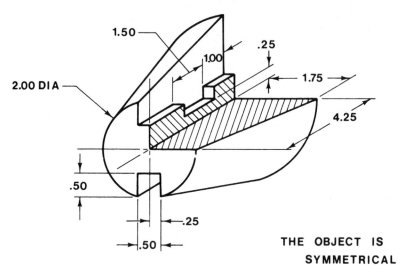

THE OBJECT IS
SYMMETRICAL

FIGURE P10-13

FIGURE P10-14

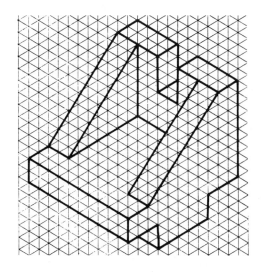

P10-15   Draw a front and side view of the object shown in Figure P10-15. Then draw two sectional views, one as defined by cutting plane A and one as defined by cutting plane B. Cutting plane A is located 1.25 from the left end of the object. Cutting plane B is located 1.00 from the right end.

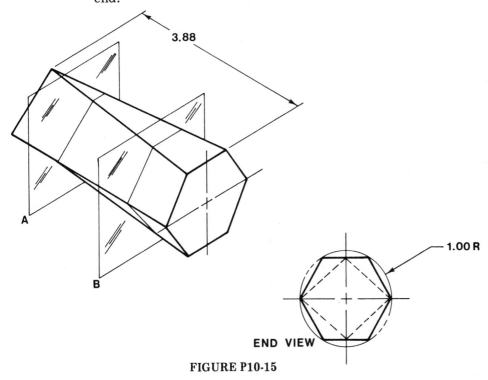

FIGURE P10-15

P10-16   Redraw the front view shown in Figure P10-16 and add a sectional view as indicated. All dimensions are in millimeters.

FIGURE P10-16

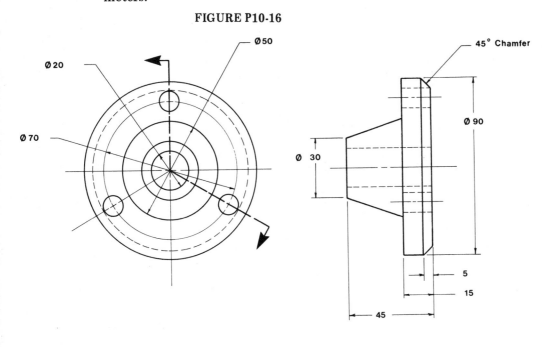

P10-17   Redraw the front view and replace the side view with a
and P10-18   sectional view taken through the vertical centerline of the
front view. All dimensions are in millimeters.

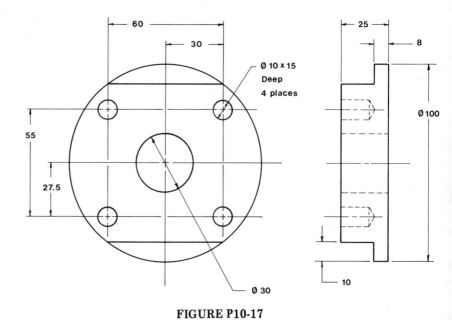

FIGURE P10-17

FIGURE P10-18

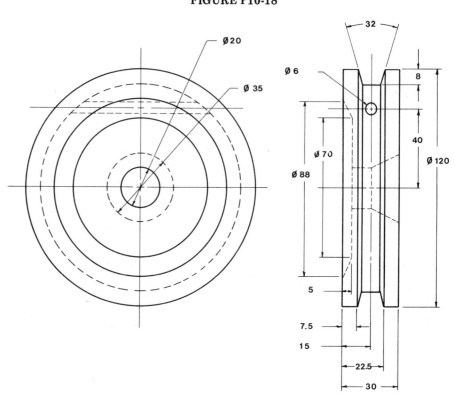

P10-19   Draw a front and a sectional view from the information
and P10-20   given. All dimensions are in millimeters.

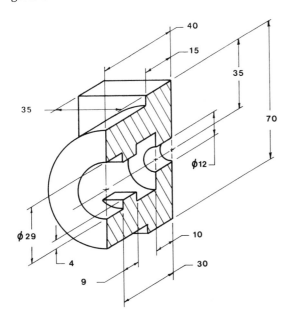

FIGURE P10-19

FIGURE P10-20

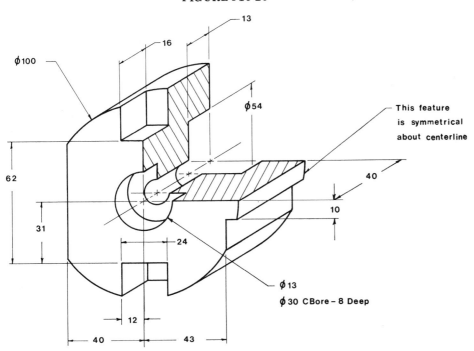

P10-21   Draw at least two orthographic views of the object shown
through   and add sectional views as indicated. Use 11 × 17 paper.
P10-23

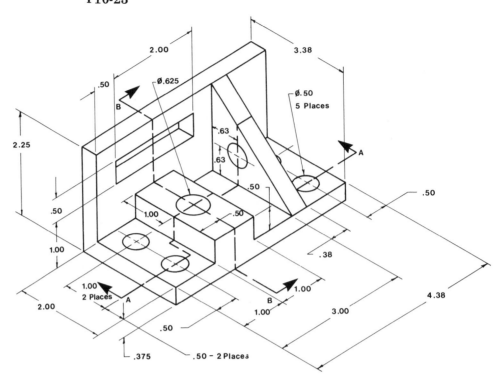

FIGURE P10-21

FIGURE P10-22   All dimensions are in millimeters.

FIGURE P10-23   All dimensions are in millimeters.

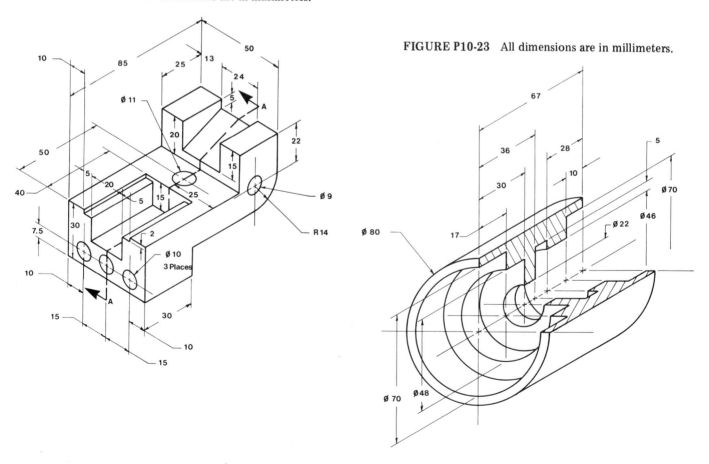

P10-24 Redraw the views given and add sectional views as indi-
and **P10-25** cated.

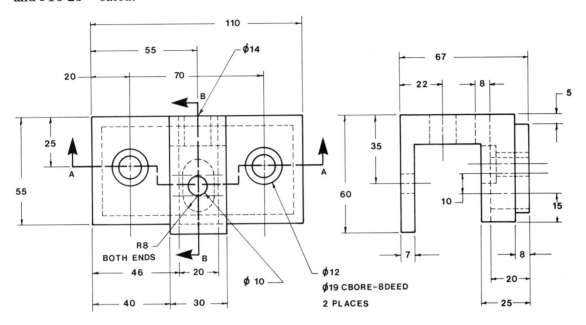

FIGURE P10-24   All dimensions are in millimeters.

FIGURE P10-25

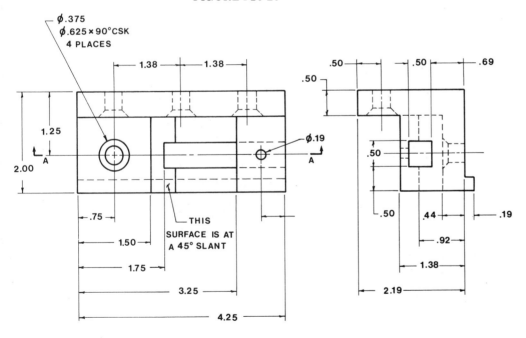

11

AUXILIARY VIEWS

FIGURE 11-0

11-1 INTRODUCTION

Auxiliary views are any orthographic views other than the three principal views. They are usually drawn to show the true shape of a surface that otherwise would appear distorted in the normal front, top, and right-side view format. For example, in Figure 11-1 neither the true shape of surface 1-2-3-4 nor the true shape of the 1/2-inch-diameter hole is shown in any of the views given. This means that even though three views of the object are presented, from a *visual* standpoint, the drawing is incomplete and therefore unsatisfactory.

Figure 11-2 shows the same object that was shown in Figure 11-1, but this time using only two orthographic views: a front view and an auxiliary view. The auxiliary view is an orthographic view taken perpendicular to surface 1-2-3-4. This two-view drawing is actually more effective in its presentation of the object than is the three-view drawing. Thanks to the auxiliary view, it defines the true shape of surface 1-2-3-4 and the 1/2-inch-diameter hole as well as all other necessary information.

Deciding when and where to use auxiliary views depends on the object being presented and on how its individual surfaces are positioned. Always use auxiliary views to add clarity to your drawings and thereby to make the technical information you are presenting easier to understand.

FIGURE 11-1   Three views of an inclined surface.

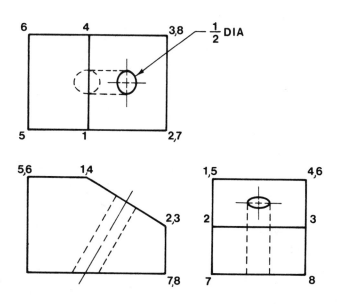

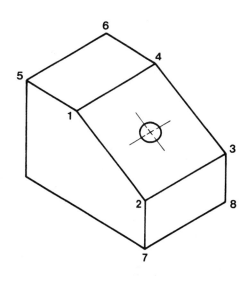

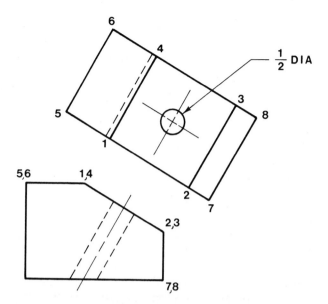

FIGURE 11-2 Front view and an auxiliary view of the object presented in Figure 11-1.

11-2   REFERENCE LINE METHOD

Two methods may be used to create auxiliary views: the reference line method, explained in this section, and the projection theory method, explained in the next section. Figure 11-3 is a sample problem in which you are given two views and are asked to create an auxiliary view that clearly presents surface 1-2-3-4-5. Figure 11-4 is the solution and was derived by using the reference line method as follows:

GIVEN: Front and side views.
PROBLEM 1: Draw an auxiliary view using the reference line method.
PROBLEM 2: Draw an auxiliary view using the projection theory method.

FIGURE 11-3

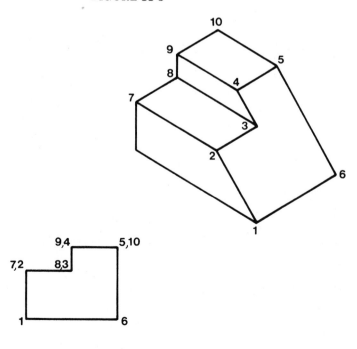

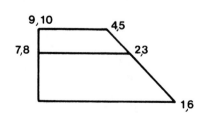

SOLUTION: Using the reference line method:

1. Draw a vertical line between the front and right-side views and draw a line parallel to surface 1-2-3-4-5-6. Define the vertical line as reference line 1. Define the line parallel to surface 1-2-3-4-5 as reference line 2.

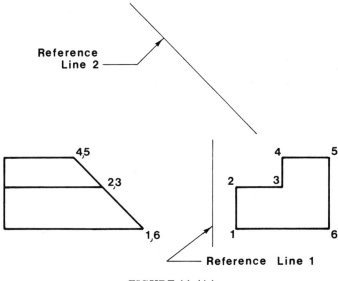

FIGURE 11-4(a)

2. Label points 1, 2, 3, 4, 5, 6, and any other points you feel you will need in both the front and right-side views.

3. Project all points in the front view into the auxiliary views by drawing very light lines perpendicular to reference line 2 from the front view points into the area where the auxiliary view will be.

FIGURE 11-4(b)

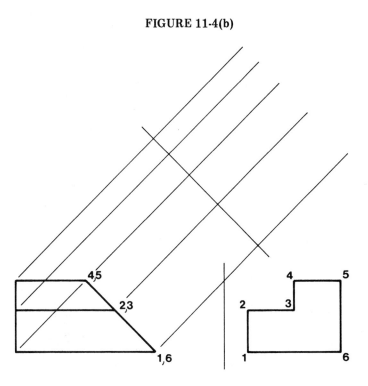

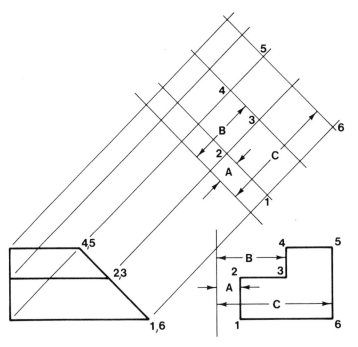

FIGURE 11-4(c)

4. Using either dividers or a compass, transfer the points from the right-side view to the auxiliary view by transferring the perpendicular distance from reference line 1 to the point, to reference line 2 along the appropriate point projection line created in step 3. This is possible because the distance between reference line 1 and the right-side view and the points is the same as the distance between reference line 2 and the auxiliary view. Label all points in the auxiliary view.

5. Lightly draw in the auxiliary view by lightly connecting the appropriate points. Check your work.

6. Erase all excess lines, point labels, and smudges and draw in all lines to their final configuration and color.

FIGURE 11-4(d)

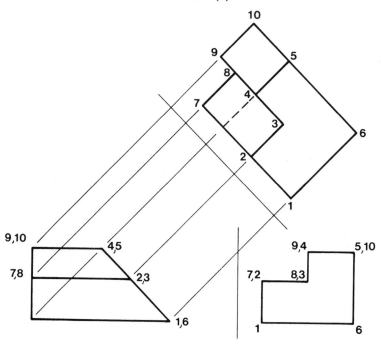

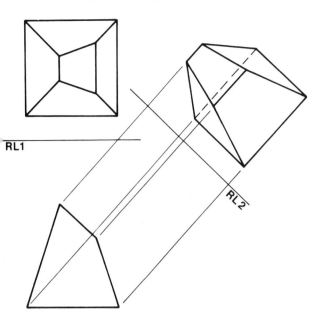

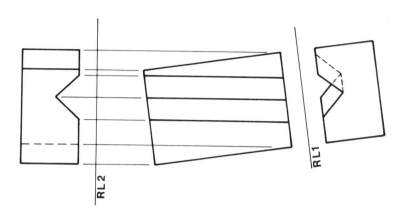

FIGURE 11-5 Auxiliary view created using the reference line method.

FIGURE 11-6 Auxiliary view created using the reference line method.

Note that surfaces 4-5-9-10 and 2-3-8-7 are distorted in the auxiliary view.

Figures 11-5 and 11-6 are further examples of auxiliary views drawn by using the reference line method.

11-3 PROJECTION THEORY METHOD

The projection theory presented in Chapter 4 may also be applied to auxiliary views. The problem of Figure 11-3 is again presented, but this time it is solved by using the projection theory method. The solution is illustrated in Figure 11-7 and was derived by using the following procedure:

SOLUTION: Using the projection method:

1. Draw a vertical line between the front and right-side views and draw a line parallel to surface 1-2-3-4-5-6. Draw the lines so that they intersect. Label the intersection point 0. Through point 0 draw two more lines: one perpendicular to the vertical line (therefore a horizontal line) and one perpendicular to the line drawn parallel to surface 1-2-3-4-5-6.

2. Project the points labeled in the side view into the area where the auxiliary view will be by first drawing vertical projection lines from the points to the horizontal line drawn in step 1. Then, using a compass set on point 0, draw projection arcs which will continue the vertical projection lines from the horizontal line to the line perpendicular to the line parallel to surface 1-2-3-4-5-6. Continue the projection lines parallel to surface 1-2-3-4-5-6 as shown.

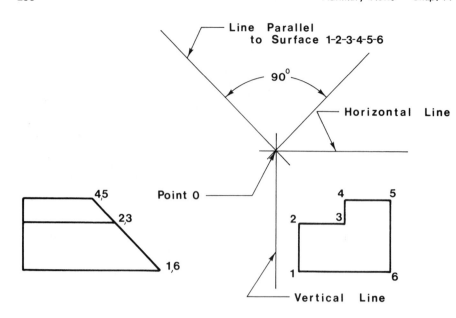

FIGURE 11-7(a)

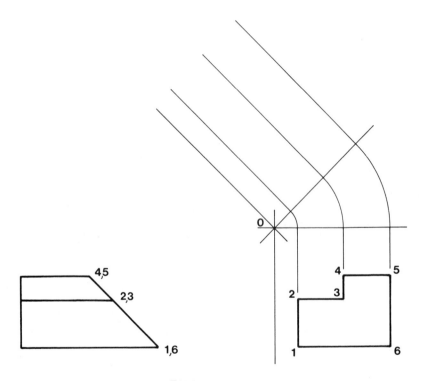

FIGURE 11-7(b)

3. Project the points from the front view by drawing lines perpendicular to the line drawn parallel to surface 1-2-3-4-5-6. Label the intersections of these projection lines with the ones drawn in step 2 with the appropriate numbers.

4. Erase all excess lines, point labels, and smudges and draw in all lines to their final configuration and color.

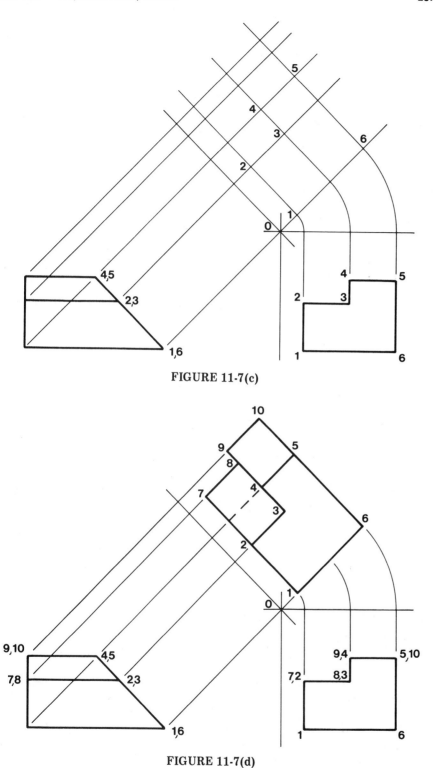

FIGURE 11-7(c)

FIGURE 11-7(d)

Note that the solution derived by the projection theory method is exactly the same as the solution derived by the reference line method. Either method will generate an accurate answer and the choice of which method to use depends on the preference of the individual drafter.

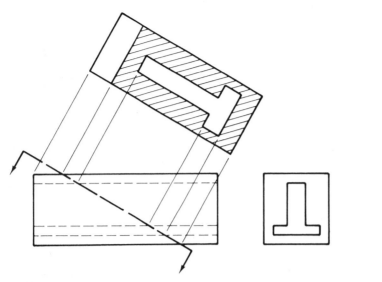

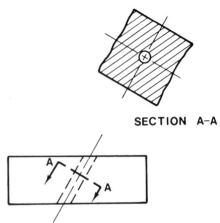

SECTION A-A

FIGURE 11-8   Auxiliary sectional view.

FIGURE 11-9   Partial auxiliary sectional view.

11-4   AUXILIARY SECTIONAL VIEWS

Auxiliary sectional views are a combination of an auxiliary view and a sectional view. They are orthographic views taken through an object at an angle defined by a cutting plane line. They adhere to the same rules and format given for sectional views in Chapter 10, and they are drawn for the same reasons: to expose surfaces that are hidden from direct view in the regular front, top, and right-side views.

Figure 11-8 is an example of a drawing that contains an auxiliary sectional view. Figure 11-9 is an example of a drawing that contains a partial auxiliary sectional view. Either the reference line or projection line method may be used to create auxiliary sectional views.

11-5   PARTIAL AUXILIARY VIEWS

Auxiliary views are helpful in clarifying drawings, but their use does have drawbacks. For example, surface 1-4-6-5, which appeared true size in the top view of Figure 11-1, appears distorted in the auxiliary view in Figure 11-3. The same is true of surface 2-7-8-3. By trying to create a view that will clarify one surface, we have distorted two other views. To eliminate distortion in the principal views, we have created auxiliary views, which in turn have created other distortions.

The solution is to use partial auxiliary views. Figure 11-10 shows the same object that was shown in Figures 11-1 and 11-2 and was drawn using a front view and a partial auxiliary view. As the name implies, a partial auxiliary view is only part of a complete auxiliary view. Partial auxiliary views enable you to limit your auxiliary view to one specific surface or part of a surface, thereby eliminating the need to draw surfaces that have become distorted in the auxiliary views.

If only one complete surface is shown in the partial auxiliary views,

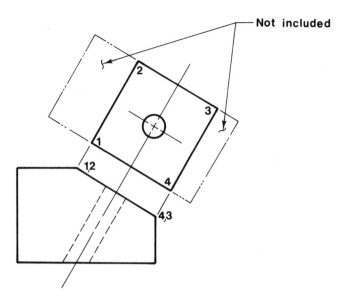

FIGURE 11-10   Front and partial auxiliary views of an object.

as is the case in Figure 11-10, break lines need not be shown. If however, part of a surface or more than one surface is to be drawn, break lines are shown.

11-6   SECONDARY AUXILIARY VIEWS

It is sometimes necessary to draw an auxiliary view of an auxiliary view. This occurs when the first auxiliary view does not completely define or does not clearly present the surface being studied. For example, Figure 11-11 shows a front, top, right-side, and auxiliary view

FIGURE 11-11   What is the true shape of surface 1-2-3? None of these views define it.

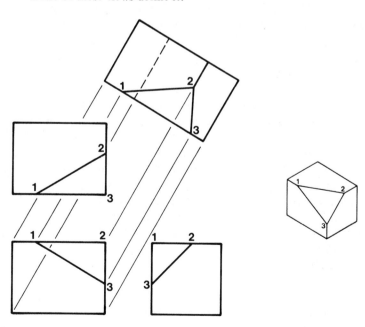

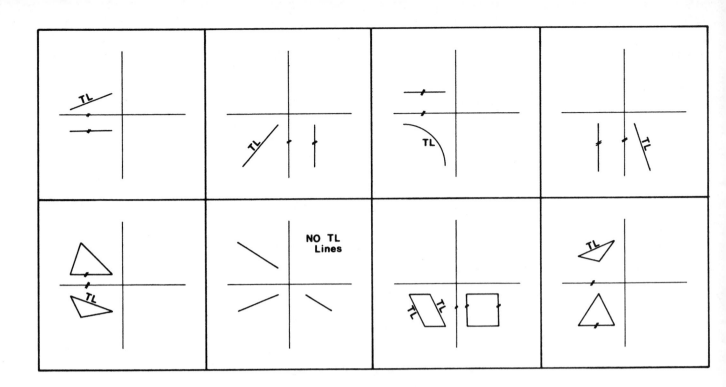

∕ Indicates lines which are Parallel

FIGURE 11-12 How to identify true length lines.

of an object that contains an oblique surface (surface 1-2-3). Despite the great number of views taken of the object, none of the given views shows the true shape of surface 1-2-3. To present a true shape of surface 1-2-3, we must use a secondary auxiliary view.

The true shape of surface 1-2-3 will only be shown in an orthographic view taken at exactly 90° to the surface. To help you visualize this concept, think of an airplane in flight. If the airplane is flying directly away from you, parallel to your line of sight (0°), it will give little or no indication of its true speed. It will simply seem to slowly disappear. If, however, the airplane is flying directly across your line of vision (90°), it will give a correct indication of its true speed. Similarly, only when a line or a plane is directly across your line of vision (an orthographic view taken at exactly 90° to your line of vision) can you see its true shape.

But how can we be assured that our secondary auxiliary view is taken at 90° to the surface? If we want our secondary view to be 90° to a surface, the first auxiliary view must be taken 0° to the surface because each auxiliary view will be 90° to the previous one.

To draw an auxiliary view that is at 0° to the surface (an end view of the surface), we must identify a true length line on the surface. A true length line is the only line on the surface whose angle relative to the principal plane lines we know exactly. Because we know the exact angle of a true length line relative to the principal plane lines, we know the angle at which to draw an auxiliary view which will be an end view of the line and therefore an end view of the surface in which the line is located.

A true length line is found by the following axiom:

An orthographic view of a line shows the true length (TL) of that line if one of the other orthographic views of that line is parallel to one of the principal plane lines. Axiom 11-1

270

Figure 11-12 illustrates this axiom. Note that as long as one of the given orthographic views is parallel to one of the principal plane lines, the other view of the line is a true length. If none of the views of the line is parallel to either principal plane line, then none of the given views is a true length. Also note that because one of the lines in a surface is true length, it does not mean that all the other lines in the surface are true length.

In the example of Figure 11-11, line 1-2 is true length in the top view, line 1-3 is true length in the front view, and line 2-3 is true length in the right-side view. We could use any one of these lines to generate an auxiliary view that is 0° to the surface 1-2-3. Line 1-2 was used for this example.

Figure 11-13 is the solution to the problem presented in Figure 11-11 and was derived by using the following procedure:

SOLUTION:

1. Identify in one of the given views a true length line. In this example line 1-2 meets the criterion set by Axiom 11-1.

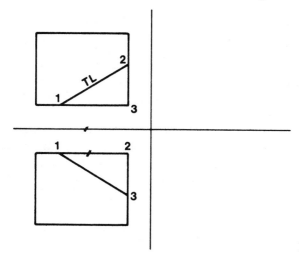

FIGURE 11-13(a)

2. Extend line 1-2 and draw lines parallel to the extension of line 1-2 throughout the other known points on the surface.

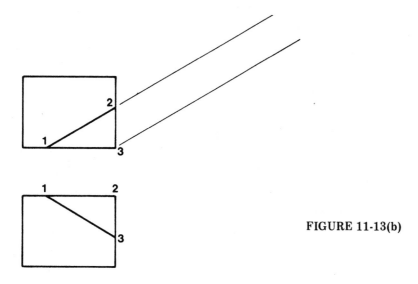

FIGURE 11-13(b)

3. Draw in the principal plane line between the two views given
    and label it reference line 1. Also draw a line somewhere along
    the extension lines drawn in step 2. The line must be perpen-
    dicular to those lines. Label it reference line 2.

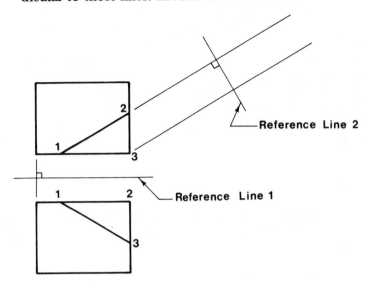

FIGURE 11-13(c)

4. Measure the distance from reference line 1 to point 1 in the
    view that contains line 1-2 parallel to the principal plane line.
    Transfer this distance to reference line 2 as shown. Make sure
    that you transfer the distance to the line that was originally
    extended through point 1. Do the same with all other points
    in the surface. Measure the distance from the point to refer-
    ence line 1, and then transfer this distance to reference line 2
    as shown.

FIGURE 11-13(d)

5. Using appropriate point numbers, label the first auxiliary view that you have now generated.

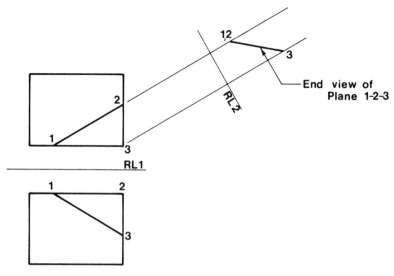

FIGURE 11-13(e)

6. Draw lines perpendicular to the auxiliary view through all points on the surface as shown. Draw a line parallel to the end view of the surface and label it reference line 3.

FIGURE 11-13(f)

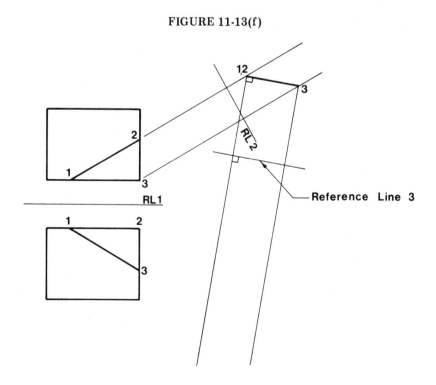

7. Measure the distance from reference line 2 to point 1 in the view in which line 1-2 appeared true length. Transfer this distance to reference line 3 as shown. Do the same with all other points in the surface.

8. Label the secondary auxiliary view of the surface with the appropriate point numbers and darken in all lines to the final color and configuration. Leave on all construction lines unless you are specifically told to erase them. This will make it easier for someone to check or follow your work.

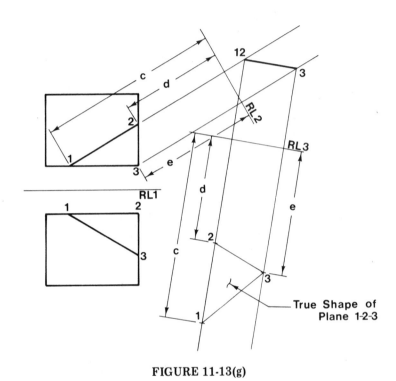

FIGURE 11-13(g)

Figure 11-14 is the solution to the problem stated in Figure 11-11, except that in Figure 11-14 line 2-3 was used to generate the first auxiliary view. The problem was solved by using the procedure outlined for Figure 11-13. Note that the true shape of surface 1-2-3 is exactly the same as that generated in Figure 11-13. Study and carefully verify how each point was transferred from reference line to reference line.

Sometimes none of the given lines that define a surface is of true length. This does not mean that a secondary auxiliary view of the surface cannot be created. It simply means that we have to create a true length line from the given information and then proceed as before. For example, the surface 1-2-3 pictured in Figure 11-15 contains no true length lines and yet the problem asks us to find the true shape of that surface which we know can only be accomplished through a secondary auxiliary view.

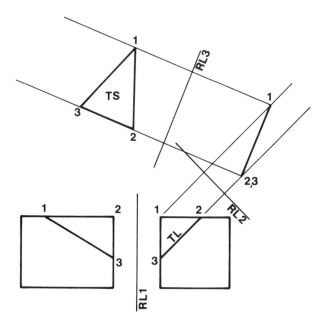

FIGURE 11-14 True shape of plane 1–2–3 found by using the true length view of line 2–3.

GIVEN: Front and top views.
PROBLEM: Draw the true shape of plane 1-2-3.

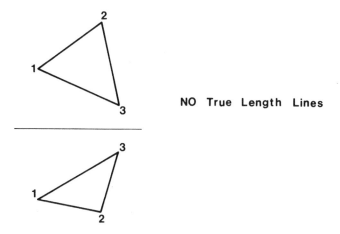

NO True Length Lines

FIGURE 11-15

To create a true length line in surface 1-2-3, first draw a line in one of the views that is parallel to one of the principal plane lines. Then project this line into the other view of the surface. In this example the new line was labeled 1-x, where point x lies along the known line 2-3. To project point x from the top view into the front view, draw a line parallel to the line drawn between the two known point 1's and perpendicular to the principal plane line from point x in the top view to a point of intersection with line 2-3 in the front view. The solution to the problem is completed as previously outlined based on the true length line 0-x. Figure 11-16 is the solution to the problem stated in Figure 11-15.

SOLUTION:

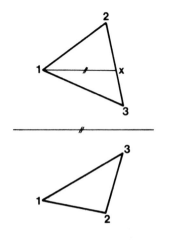
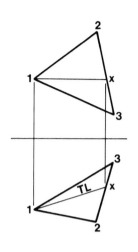
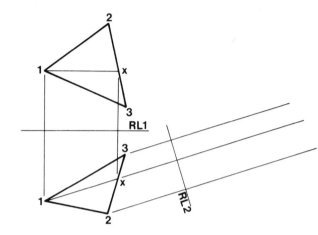
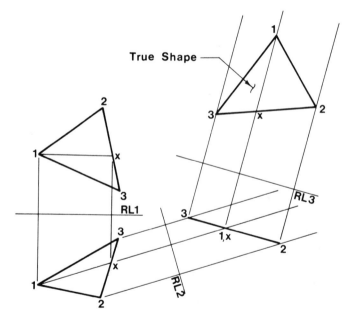

FIGURE 11-16

PROBLEMS

P11-1   Redraw each object and add the appropriate auxiliary
through views.
P11-11

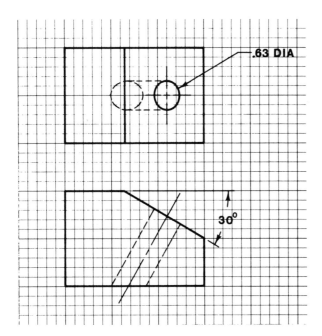

FIGURE P11-1

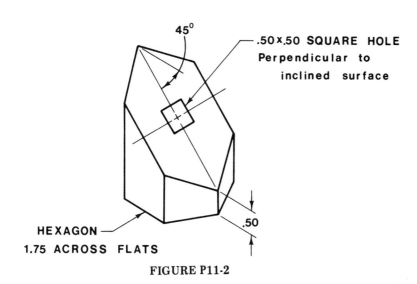

45°

.50 × .50 SQUARE HOLE
Perpendicular to
inclined surface

HEXAGON
1.75 ACROSS FLATS

.50

FIGURE P11-2

FIGURE P11-3

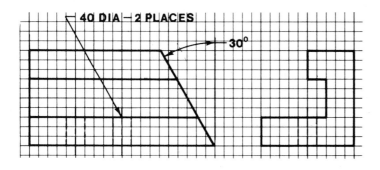

40 DIA — 2 PLACES

30°

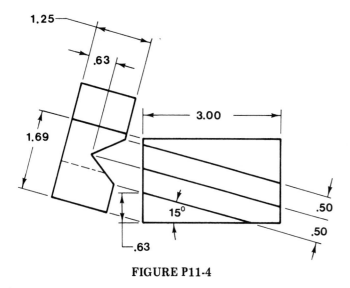

FIGURE P11-4

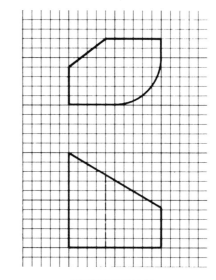

FIGURE P11-5

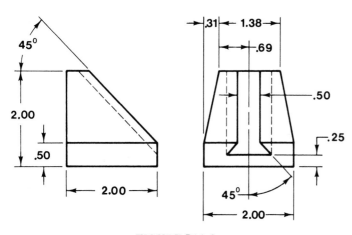

FIGURE P11-6

FIGURE P11-7

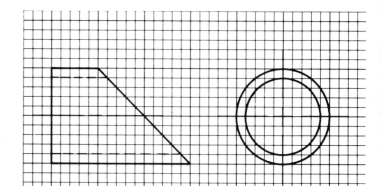

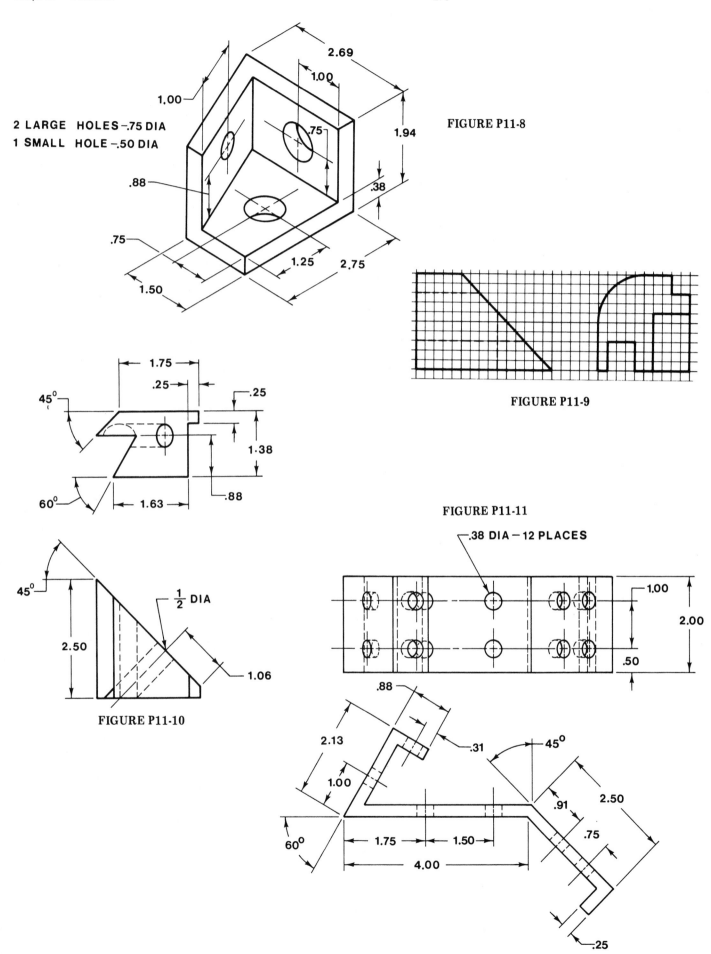

2 LARGE HOLES –.75 DIA
1 SMALL HOLE –.50 DIA

FIGURE P11-8

FIGURE P11-9

FIGURE P11-10

FIGURE P11-11

.38 DIA – 12 PLACES

P11-12   Using a secondary auxiliary view, derive the true shape of
and P11-13   each plane. Each square of the grid is 0.20 per side.

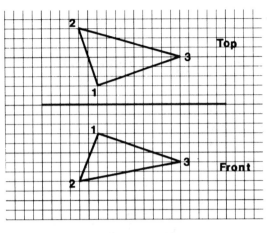

FIGURE P11-12

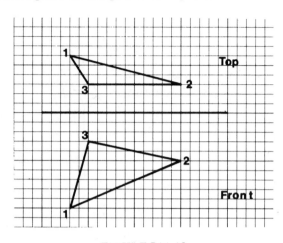

FIGURE P11-13

P11-14   Draw sufficient views to completely define each object.
through   In each case, include a secondary auxiliary view of the
P11-17   oblique surface.

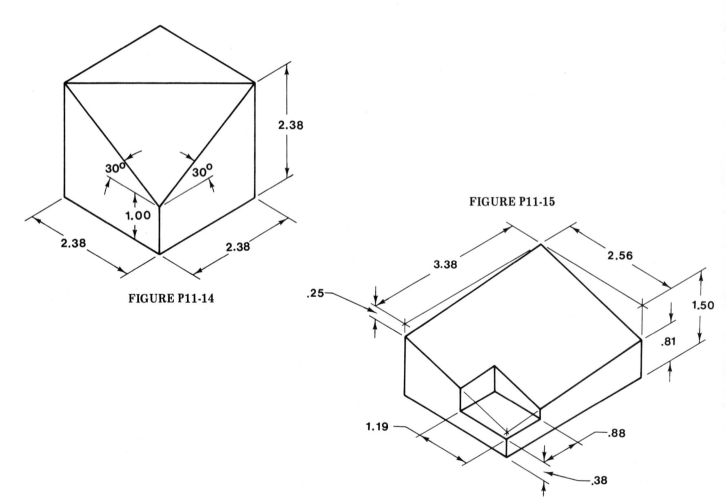

FIGURE P11-15

FIGURE P11-14

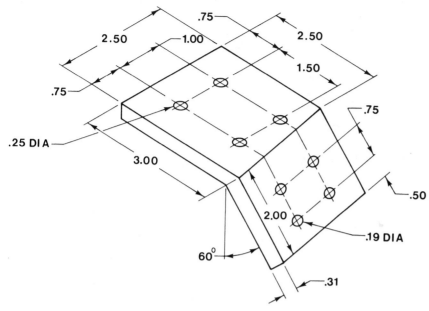

FIGURE P11-16

FIGURE P11-17

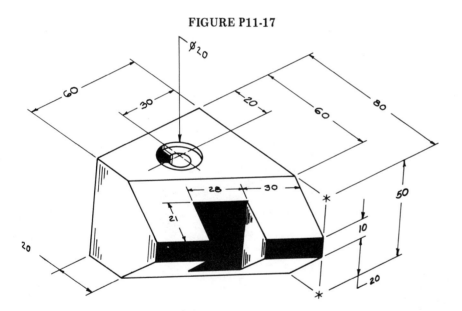

12
FASTENERS

FIGURE 12-0

12-1   INTRODUCTION

There are two basic fasteners: mechanical and nonmechanical. Mechanical fasteners include bolts, rivets, and screws and, from a design standpoint, they are usually stronger, easier to work with, and more easily replaced than nonmechanical fasteners. Nonmechanical fasteners include glues, epoxies, tapes, and so on, and they are usually less expensive, lighter, and require less installed space than do mechanical fasteners.

This chapter deals exclusively with mechanical fasteners. Nonmechanical fasteners are not drawn, but they are noted on a drawing as shown in Figure 12-1. Mechanical fasteners, however, have specific representations that must be clearly and accurately drawn.

12-2   THREAD TERMINOLOGY

Figure 12-2 illustrates some of the basic terms used to describe a thread. These terms are common to all kinds of threads and will be referred to throughout the chapter.

The pitch of a thread is equal to 1 over the number of threads per inch.

$$P = \frac{1}{\text{number of threads per inch}} \qquad (12\text{-}1)$$

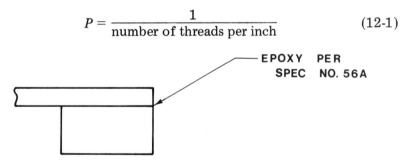

FIGURE 12-1   Callout for a nonmechanical fastener.

FIGURE 12-2   Some basic terms used to describe a thread.

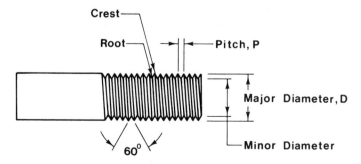

A thread made with 20 threads per inch, for example, has a pitch of 0.05 inch.

$$P = \frac{1}{20} = 0.05 \text{ inch}$$

A thread with eight threads per inch has a pitch of 0.125 inch.

$$P = \frac{1}{8} = 0.125 \text{ inch}$$

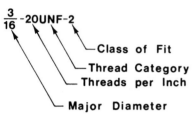

FIGURE 12-3 Definition of a thread notation.

12-3   THREAD NOTATIONS

Figure 12-3 shows a typical thread notation and a definition of each term. The terms *major diameter* and *threads per inch* (pitch) have already been explained in Section 12-2 and Figure 12-2. The terms *thread category* and *class of fit* require further explanation.

Threads are generally manufactured to either National Coarse or National Fine standards, although there are several other categories of thread standards (Unified Extra Fine, for example). These standards are internationally agreed upon manufacturing specifications that result in products of uniform quality and interchangeability. From a drawing standpoint, there is no difference between any of the standards, for they all use the same representations.

Class of fit refers to the way in which two threads match each other. There are four categories: classes 1, 2, 3, and 4. The higher the number, the better quality the matchup. Class 1 is a very sloppy fit; class 2 is the most commonly manufactured fit and is generally acceptable in most design situations; classes 3 and 4 are rarely specified because they are very exact and very expensive.

Figure 12-4 shows a metric thread designation. The letter M always precedes the nominal diameter. In Figure 12-4, the nominal thread

FIGURE 12-4 A sample metric thread callout.

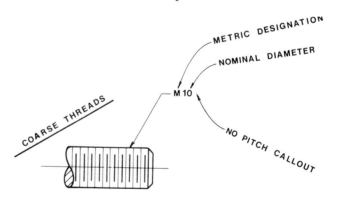

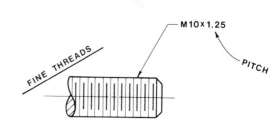

diameter equals 10 millimeters. Coarse thread size requires only the M and the nominal thread size. No pitch size is needed. Fine threads *do* require a pitch size callout. The pitch size callout is listed after the thread size and is separated from it by an X. Common metric thread sizes are listed in Appendix D.

12-4   THREAD REPRESENTATION

There are three ways to represent threads on a drawing: detailed, schematic, and simplified. From a drawing standpoint, each representation has advantages and disadvantages. The detailed representation is very easy for the reader to understand, but it is very time consuming to draw. The simplified representation is very easy to draw, but to the uneducated reader, it is very confusing. The schematic representation is a compromise—fairly easy to draw and fairly easy to read, but it is still inexact and time consuming. The representation chosen will depend on the specific shop or drafting requirements applicable.

Simplified Representation (Figure 12-5)

(a) Define the major diameter, thread length, and shaft length of the desired thread.

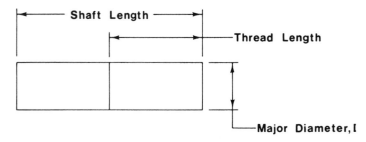

FIGURE 12-5(a)   Simplified thread representation.

(b) Draw a 45° chamfer 1/16 or 1/8 long on the end of the threaded portion of the shaft. The choice of 1/16 or 1/8 depends on which looks better.

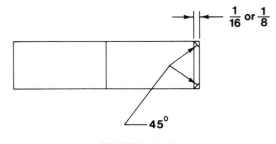

FIGURE 12-5(b)

(c) Draw hidden lines as shown. If you used 1/16 in step (b), then use it here. If you used 1/8, then use 1/8.

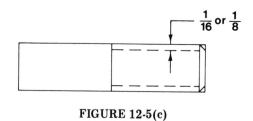

FIGURE 12-5(c)

(d) Darken in the visible lines and add the appropriate thread call-out.

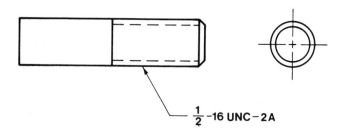

$\frac{1}{2}$-16 UNC-2A

FIGURE 12-5(d)

Schematic Representation (Figure 12-6)

(a) Define the major diameter, thread length, and shaft length of the desired thread.

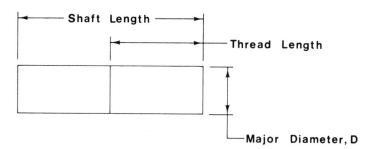

FIGURE 12-6(a)   Schematic thread representation.

(b) Draw parallel lines as shown. Draw these lines extremely lightly because they will be erased later. The choice of 1/16 or 1/8 depends on which looks better.

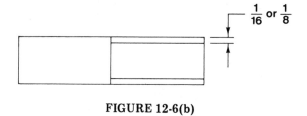

FIGURE 12-6(b)

(c) Draw lines perpendicular to the lines drawn in step (b) as shown. If 1/16 was used in step (b), space them 1/16 apart. If 1/8 was used in step (b), space them 1/8 apart.

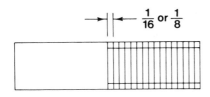

FIGURE 12-6(c)

(d) Draw 45° chamfers at the threaded end of the shaft as shown. Draw lines parallel to and halfway between the lines drawn in step (c). Start and end these lines as they intersect the lines drawn in step (b).

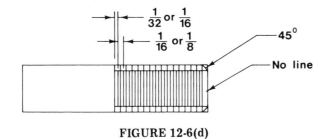

FIGURE 12-6(d)

(e) Darken the lines created in step (d) and all visible lines as shown. Add the appropriate thread callout.

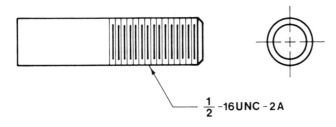

FIGURE 12-6(e)

If desired, the spacing of the lines drawn in step (c) may be made exactly equal to the thread pitch.

Detailed Representation (Figure 12-7)

(a) Define the major diameter, thread length, and shaft length of the desired thread.

FIGURE 12-7(a)   Detailed thread representation.

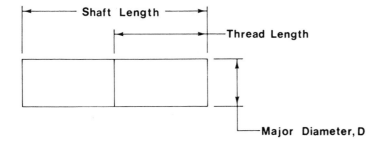

(b) Along the top edge of the shaft mark off as many distances *P* as will fit within the desired thread length. Mark off a distance of 1/2*P* along the bottom edge.

Note: This is a right-hand thread. When the designated thread is a left-hand thread, the *P* distances would be marked off along the lower edge and the 1/2*P* distance along the top edge.

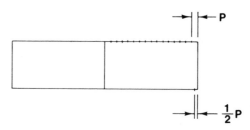

FIGURE 12-7(b)

(c) Connect the first *P* distance with the 1/2*P* distance. Then draw lines, parallel to this line, through each of the *P* distances as shown.

FIGURE 12-7(c)

(d) Draw short 60° lines as shown.

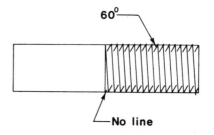

FIGURE 12-7(d)

(e) Draw short 60° lines so that they intersect the lines drawn in step (d) as shown.

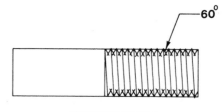

FIGURE 12-7(e)

(f) Connect the intersections of the 60° lines as shown. These lines are not parallel to the lines drawn in step (c).

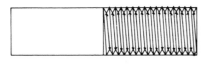

FIGURE 12-7(f)

(g) Darken the lines as shown and add the appropriate thread callout.

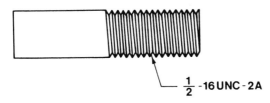

$\frac{1}{2}$ -16 UNC - 2A

FIGURE 12-7(g)

12-5   THREADS IN A SECTIONAL VIEW

Figure 12-8 shows the three different thread representations as they appear in a sectional view. Note that the simplified representation includes hidden lines. Hidden lines are drawn in sectional views when a simplified representation is used and in all end views of threaded holes regardless of the representation.

FIGURE 12-8   (a) Simplified representation; (b) schematic representation; (c) detailed representation.

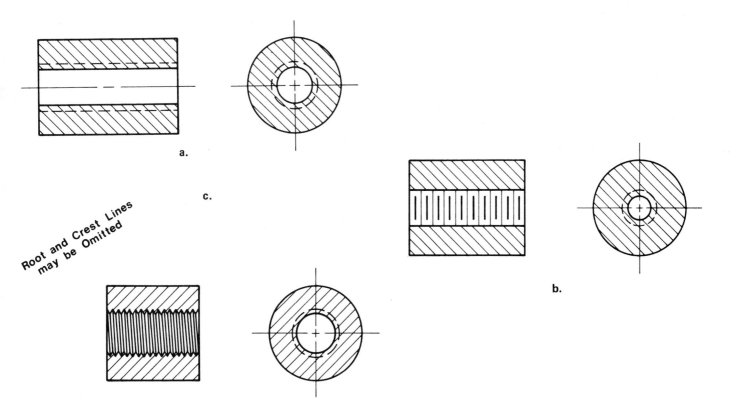

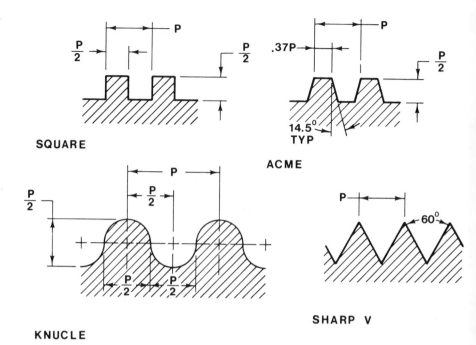

FIGURE 12-9   Various thread profiles.

12-6   THREADS

There are several different kinds of threads: square, acme, knuckles, sharp V, and others. Figure 12-9 shows profiles of these threads.

A double thread has two threads cut on the same shaft. When it is rotated, it advances or recedes twice as fast as a single thread (one revolution of a double thread will transverse twice the distance traveled by one revolution of a single thread). Double threads may be cut in any thread—square, UNC, UNF, and so on. Figure 12-10 includes a double thread drawn by using simplified representation. Note how the thread note is written and that the picture portion of the drawing is the same as for single threads.

Most threads are right-hand threads—that is, they advance when they are turned clockwise. There are also left-hand threads. The oxygen lines in most hospitals are made with left-hand threads as a safety precaution to prevent an accidental mix-up with other gas lines. The schematic and detailed representations are drawn the same for left- or right-hand threads. Only the notation is amended to include an "LH" for left-hand threads. It is assumed that a thread is a right-hand thread if LH does not appear. Figure 12-11 illustrates a callout note for a left-hand thread.

FIGURE 12-10   Double thread callout and simplified representation.

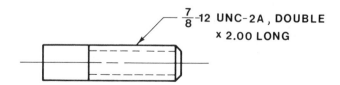

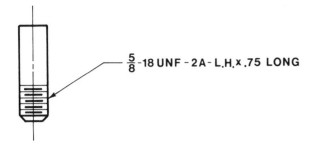

$\frac{5}{8}$-18 UNF - 2A - L.H. x .75 LONG

FIGURE 12-11 Left-hand thread callout and schematic representation.

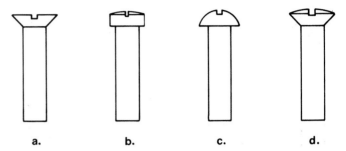

FIGURE 12-12 Different types of mechanical fasteners: (a) flathead; (b) fillester head; (c) round head; (d) oval head.

The detailed representation of a left-hand thread is different from the detailed representation of a right-hand thread. To draw a left-hand thread, use the same procedure but change the initial $P/2$ offset shown in step (b) of Figure 12-7 from the top edge to the bottom edge of the thread.

12-7   TYPES OF BOLTS AND SCREWS

Figure 12-12 illustrates several of the many different mechanical fasteners that are commercially available. The exact size and shape specifications are available from the manufacturers.

12-8   THREADED HOLES

When you draw a threaded hole representation, it is important to know how such a hole is created. First, a hole, called a *pilot hole*, is drilled. This hole is then tapped (threads are cut into the surface of the pilot hole). Holes are not usually tapped all the way to the bottom of the pilot hole because this would cause severe damage to the tapping bit (although special tapping bits are available that will tap to the bottom of a pilot hole).

When you draw a threaded hole, always show the untapped portion of the pilot hole as illustrated in Figure 12-13. The pilot hole usually extends the equivalent of two thread lengths beyond the tapped portion of the hole. For example, if we wish to draw a threaded hole in which the thread depth is to be 3 and the thread type is to be 1-8UNC-2, we would first calculate the depth of one thread by using Equation (12-1):

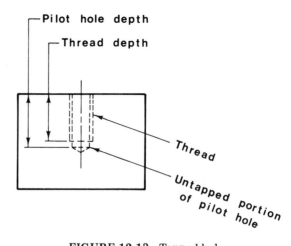

FIGURE 12-13   Tapped hole.

$$P = \frac{1}{\text{number of threads per inch}}$$

$$= \frac{1}{8}$$

$$= 0.125$$

$$2P = 0.250$$

See Figure 12-14 for an example of this. We would then calculate the depth of the pilot hole by adding the total length of the thread to the equivalent of two thread lengths.

$$\text{total thread length} + 2P = \text{pilot drill depth}$$

$$3.000 + 0.250 = \text{pilot drill depth} \qquad (12\text{-}2)$$

FIGURE 12-14   Orthographic view, a sectional view using the simplified representation, and a sectional view using the schematic representation of a 1-8UNC-2 $\times$ 3 thread.

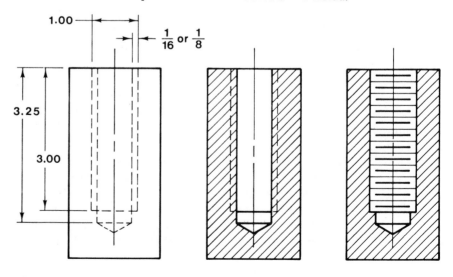

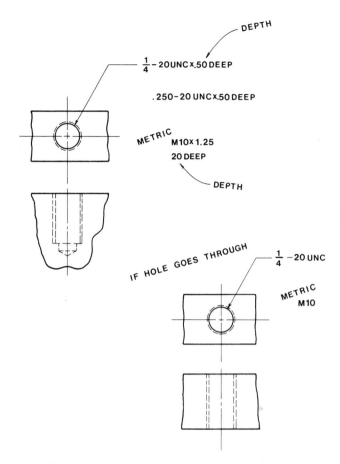

FIGURE 12-15   Sample drawing callouts for threaded holes.

Figure 12-15 shows a drawing callout for threaded holes that do not pass completely through an object. The depth required for English thread sizes is listed after the thread size callout and is separated from it by an X. The depth callout consists of two parts: the numerical value and the word DEEP.

The depth requirement for metric thread designations is listed under the thread callout as shown in Figure 12-15. If no depth callout is given, the hole is assumed to pass completely through the object. This is true for both English and metric callouts.

If we wished to draw a threaded hole with a 3/8-16UNC-2 thread cut to a depth of 1.38, the calculations would be as follows:

From Equation (12-1):

$$P = \frac{1}{16}$$

$$= 0.06$$

$$2P = 0.12$$

From Equation (12-2):

$$1.38 + 0.12 = \text{pilot drill depth}$$

$$1.50 = \text{pilot drill depth}$$

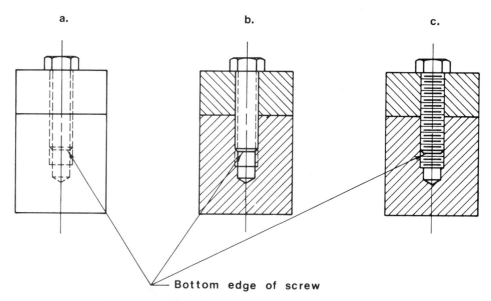

Bottom edge of screw

FIGURE 12-16 Threaded hole with a screw assembled into it: (a) orthographic view; (b) simplified representation; (c) schematic representation.

Figure 12-16 illustrates a threaded hole with a screw assembled into it. Note how the bottom of the screw is distinguished from the threads by the 45° chamfers and also note that the threads extend beyond the bottom of the screw. Threads usually extend at least two thread lengths beyond the bottom of a screw to prevent the screw from bottoming and jamming in the hole.

To draw a threaded hole with a fastener assembled in it, calculate the thread depth from Equation (12-3) and the pilot hole depth from Equation (12-2):

$$\text{threaded hole depth} = \text{fastener depth} + 2P \quad (12\text{-}3)$$

For example, to draw a threaded hole for a 3/4-10UNC-1 × 2.50 machine screw, first calculate the threaded length by using Equation (12-1):

$$P = \frac{1 \text{ inch}}{10}$$

$$= 0.10$$

$$2P = 0.20$$

Calculate the threaded hole depth by using Equation (12-3):

$$\text{threaded hole depth} = 2.50 + 0.20 = 2.70$$

Finally, calculate the pilot hole depth by using Equation (12-2):

$$\text{pilot hole depth} = 2.70 + 0.20 = 2.90$$

A table of pilot hole diameters for various thread diameters is included in Appendix D.

12-9   DRAWING BOLT AND SCREW HEADS

Figure 12-17 illustrates how to draw a hex head bolt. The procedure used is as follows:

(a) Define the diameter and length of the bolt.

(b) Draw a circle of 1-1/2D diameter as shown. Draw a line parallel to the top of the bolt shank as a distance 2/3D as shown. *Note:* The term 1-1/2D means one and one-half times the diameter; similarly for 2/3D. If, for example, the diameter of the bolt were 1/2, 1-1/2D would equal

$$1\tfrac{1}{2}D = \frac{3}{2}\left(\frac{1}{2}\right) = \frac{3}{4}$$

2/3D would equal

$$\frac{2}{3}D = \frac{2}{3}\left(\frac{1}{2}\right) = \frac{2}{6} = \frac{1}{3}$$

(c) Circumscribe a hexagon around the 1-1/2D circle.

(d) Project the hexagon's corners as shown.

(e) Draw a line 60° to the horizontal through the intersection of the outside corner projection line and 2/3D line created in step (b) such that it crosses the centerline of the bolt. Do the same for the other corner intersection. Label the intersection of the two 60° lines point 1.

(f) Draw a 60° line through each of the intersections of the inside projection lines and the 2/3D line created in step (b). Label the intersections of these 60° lines with those created in step (e) as points 2.

(g) Using point 1 and both points 2 as compass points, draw arcs as shown. Darken the appropriate lines and add the desired thread notation.

FIGURE 12-17   How to draw a hex head bolt.

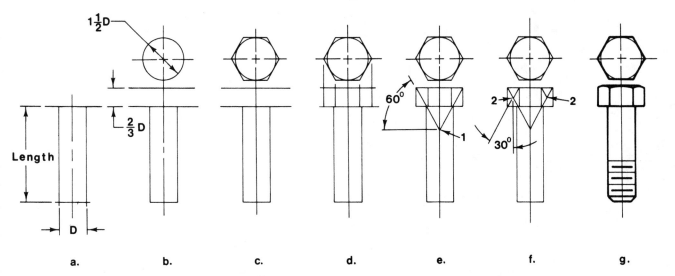

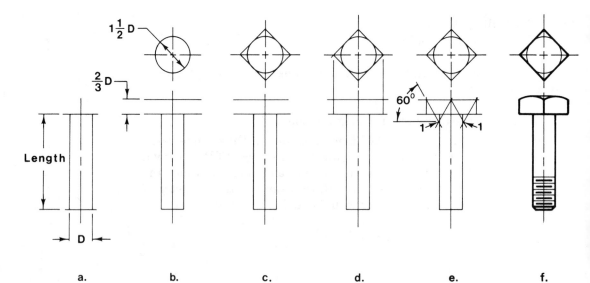

a.   b.   c.   d.   e.   f.

FIGURE 12-18   How to draw a square head bolt.

Figure 12-18 illustrates how to draw a square head bolt. The procedure used is as follows:

(a)   Define the diameter and length of the bolt.

(b)   Draw a circle of 1-1/2D diameter as shown. Draw a line parallel to the top of the bolt shank at a distance 2/3D as shown.

(c)   Circumscribe a square around the 1-1/2D circle.

(d)   Project the square's corners as shown.

(e)   Draw 60° lines through the intersection of the projection lines drawn in step (d) with the 2/3D line drawn in step (b). Label the two intersections of the 60° lines as points 1.

(f)   Using the points 1 as compass centers, draw in the arcs as shown. Darken in the appropriate lines and add the desired thread notation.

Nuts are drawn by using the same procedures as for bolt heads except that they are 7/8D high instead of 2/3D high. Figure 12-19 illustrates hex and square nuts.

FIGURE 12-19   How to draw hex and square nuts.

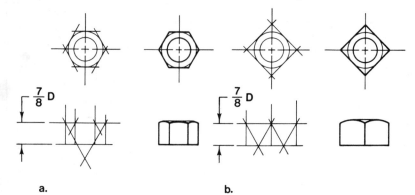

a.   b.

Detailed   Schematic

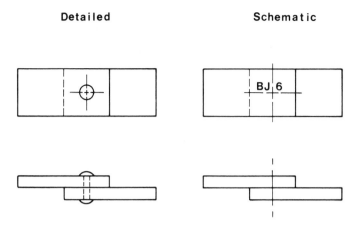

FIGURE 12-20   Rivet representations.

12-10   RIVETS

Rivets are metal fasteners that are commonly used to hold sheet metal parts together. Although they are inexpensive and lightweight, they are not as strong as screws or bolts. Rivets are not reusable and once they are placed in an assembly, they can only be removed by drilling.

Figure 12-20 illustrates two of the many representations used to call out rivets on a drawing. The detailed representation in the top view consists of circles with diameters equal to the diameter of the rivet's head. The side view is as shown. In the top view the schematic representation consists of short, perpendicularly crossed lines that locate the center of the rivet. The side view looks like a centerline of a hole, except that it always ends with a short line.

The meaning of the callouts for schematic representations is illustrated in Figure 12-21. The actual identification letter designations (BJ, CX, HY, and so on) vary from company to company, although most aircraft companies use the National Aircraft Standards (NAS).

A long row of rivets, provided that the rivets are all exactly the same kind, may be called out by calling out only the first and last rivet in the row. Figure 12-22 illustrates this kind of rivet callout.

FIGURE 12-21   Meaning of schematic representation rivet callouts.

FIGURE 12-22   How to call out rows of rivets.

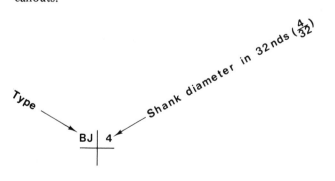

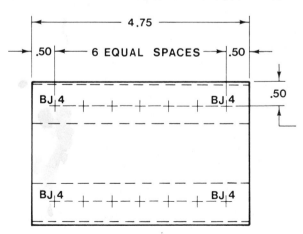

12-11 WELDS

Welds are usually called out on a drawing by notes such as are shown in Figure 12-23. There are many different welds. Interested students are referred to the American Welding Society, 2501 N.W. 7th St., Miami, FL 33125.

FIGURE 12-23 Weld callouts per standards set by the American Welding Society.

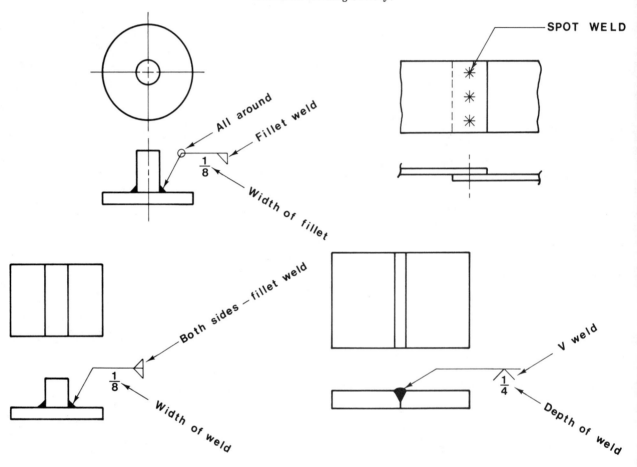

PROBLEMS

P12-1 Redraw each figure, adding the appropriate fasteners. Use the
through representation specified by your instructor. Each square on
P12-4 the grid pattern is 0.20 per side.

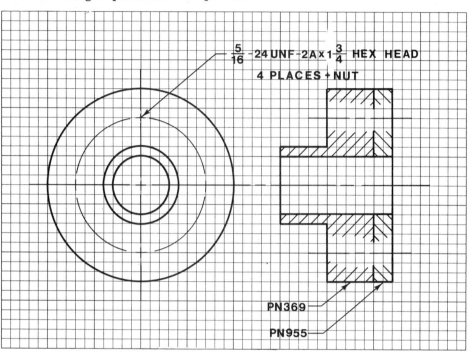

FIGURE P12-1

FIGURE P12-2

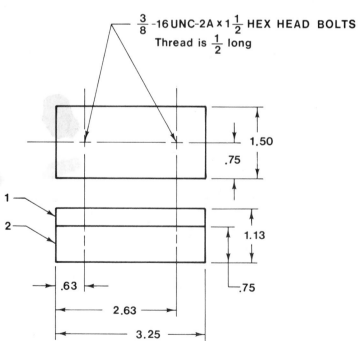

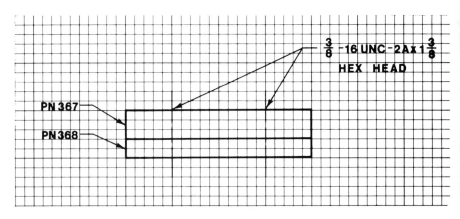

FIGURE P12-3

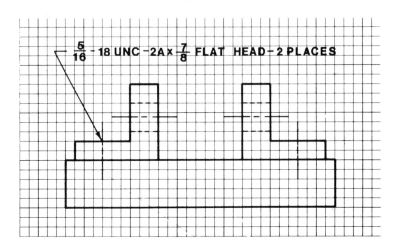

FIGURE P12-4

P12-5   Redraw the sectional view shown in Figure P12-5 and add the
following fasteners:
a. 0.312-18UNC-2A × 1.25 HEX HEAD SCREW
b. 0.438-14UNC-2A × 1.38 SQUARE HEAD SCREW
c. #10 (0.190)-32UNF × 1.50 HEX HEAD SCREW
d. 0.562-18UNF × 1.00 SQUARE HEAD SCREW

FIGURE P12-5

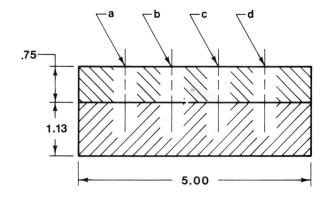

P12-6 The shop complains that the fastener callouts shown in Figure P12-6 are incorrect because the heads interfere, that is, bump into one another. Prepare a layout to verify if this is true. If it is true, how would you alleviate the interference?

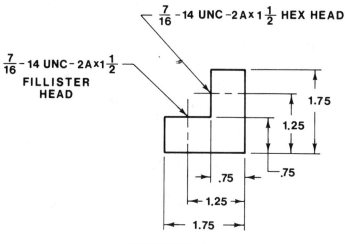

FIGURE P12-6

P12-7 Redraw Figure P12-7 and add the appropriate rivet callouts. Make the end rivets of each row (first and last rivet) BJ6s and all other rivets BJ4s.

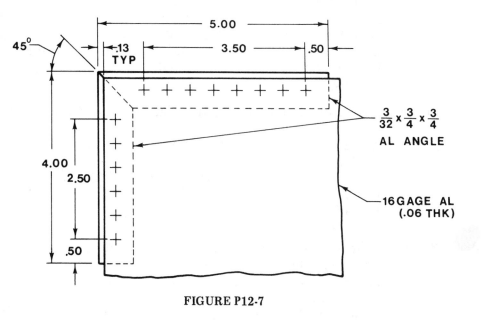

FIGURE P12-7

P12-8 Prepare a drawing of Parts 1 and 2 shown in Figure P12-8 using the dimensions listed below. Draw both objects on the same sheet of paper. Choose a fastener from Appendix D that matches the threads in Part 2. Choose a standard length for the fastener that allows for the thickness of Part 1 and does not bottom out in the hole in Part 2. Reference the

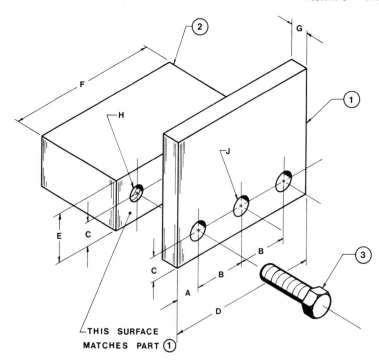

FIGURE P12-8

chosen fastener using a callout as defined in Section 12-4. All values are in inches.

A. 0.50   F. 3.00
B. 1.00   G. 0.375
C. 0.50   H. 0.312-24 UNF × 1.00 DEEP
D. 3.00   J. 0.375
E. 1.00

P12-9   Repeat Problem 12-8 using the dimensions listed below. All values are in millimeters.

A. 12.0   F. 72.0
B. 24.0   G. 10.0
C. 12.0   H. M10 × 24 DEEP
D. 72.0   J. 014
E. 24.0

P12-10   Redraw the sectional view shown in Figure P12-5 and add the following fastener:

a. M10 × 30 HEX HEAD SCREW
b. M14 × 35 SQUARE HEAD SCREW
c. M8 × 1.25 × 32 HEX HEAD SCREW
d. M20 × 2.5 × 30 SQUARE HEAD SCREW

13
METRICS

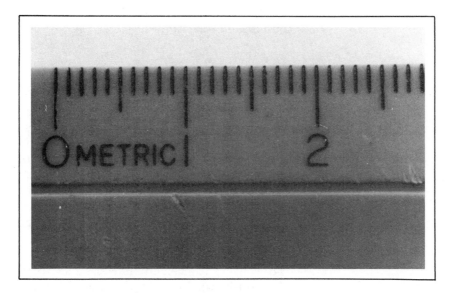

FIGURE 13-0

13-1 INTRODUCTION

Many large corporations, both in the United States and in other countries, are multinational corporations. They operate plants, buy goods, and sell products in many countries. Olivetti-Underwood, for example, is an Italian company headquartered in Ivrea, Italy, but it owns and operates manufacturing plants in Spain, the United Kingdom, the United States, Argentina, Brazil, Colombia, and Mexico, and it sells its products worldwide.

Because so many companies operate internationally, engineers and drafters must be prepared to exchange technical information internationally. This may be difficult, not only because of the language differences, but also because of the different systems used to measure and present technical information. In the United States we use the English system of measuring (feet and inches) and third-angle projections for presenting orthographic views. Most other countries use the metric system of measuring (meters and millimeters) and first-angle projections for presenting orthographic views.

Because the metric system is easier to use than the English system, all major nonmetric countries have started to change their engineering measuring systems to the metric system, but the change has not yet been completed. In the United States the change has been slow, primarily because of the enormous costs involved in replacing existing nonmetric tools and machinery. Until the metric system becomes universal, it is important that drafters know how to work comfortably in *both* systems. This chapter will explain the metric system and first-angle projection and then it will compare them with the English system and third-angle projection.

13-2 THE METRIC SYSTEM

In the metric system measurements of length are based on a fixed unit of distance called a *meter*. A meter is slightly longer than a yard. A meter is divided into smaller units called *centimeters* and *millimeters*. There are 100 centimeters or 1000 millimeters to a meter. Most mechanical measurements in the metric system are made by using millimeters just as most mechanical measurements in the English system are made by using inches.

The symbol for a millimeter is mm (5 mm, 26 mm, and so on). Figure 13-1 shows a millimeter scale together with a few sample measurements.

To convert a given millimeter value to meters, divide the given value by 1000, which is the same as shifting the decimal point three places to the left. For example,

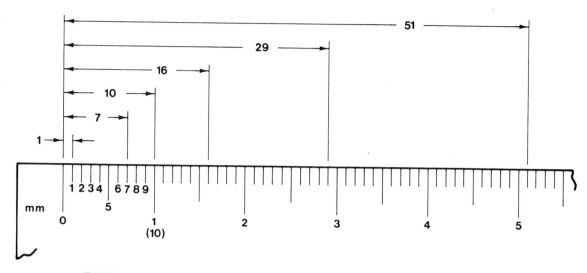

FIGURE 13-1   Millimeter scale with sample measurements.

$$423 \text{ mm} = \text{how many meters?}$$

$$\frac{423}{1000} = 0.423 \text{ m}$$

To convert a given meter value to millimeters, multiply the given value by 1000, which is the same as shifting the decimal point three places to the right. For example,

$$5.1 \text{ m} = \text{how many millimeters?}$$

$$(5.1)(1000) = 5100 \text{ mm}$$

All intermetric distance conversions are accomplished in a similar manner. Remember that there are 10 millimeters to 1 centimeter, 100 centimeters to 1 meter, and 1000 millimeters to 1 meter.

13-3 CONVERSION BETWEEN MEASURING SYSTEMS

To convert millimeters to inches or inches to millimeters, use the following equality:

$$25.4 \text{ mm} = 1 \text{ inch}$$

If you are given a value in millimeters and wish to convert it to an inch value, divide the millimeter value by 25.4. For example,

$$354 \text{ mm} = \text{how many inches?}$$

$$\frac{354}{25.4} = 13.94 \text{ inches}$$

$$10 \text{ mm} = \text{how many inches?}$$

$$\frac{10}{25.4} = 0.394 \text{ inches}$$

If you are given a value in inches and wish to convert it to a millimeter value, multiply the inch value by 25.4. For example,

$$3.20 \text{ inches} = \text{how many millimeters?}$$

$$(3.20)(25.4) = 81.28 \text{ mm}$$

$$0.68 \text{ inch} = \text{how many millimeters?}$$

$$(0.68)(25.4) = 17.27 \text{ mm}$$

If you are given a fractional inch value and wish to change it to a millimeter value, you must first change the fractional value to its decimal equivalent in inches and then multiply the decimal value by 25.4. For example,

$$6\frac{7}{8} \text{ inches} = \text{how many millimeters?}$$

$$6\frac{7}{8} \text{ inches} = 6.88 \text{ inches}$$

$$(6.88)(25.4) = 174.75 \text{ mm}$$

$$\frac{9}{16} \text{ inch} = \text{how many millimeters}$$

$$\frac{9}{16} \text{ inch} = 0.56 \text{ inch}$$

$$(0.56)(25.4) = 14.22 \text{ mm}$$

13-4 CONVERSION TABLES

This section contains two conversion tables: one for converting inches to millimeters (Table 13-1), and one for converting millimeters to inches (Table 13-2). Conversion tables enable you to convert given values directly without having to go through extensive calculations. The tables, however, are limited and any values not included in them must be converted mathematically.

To use the inches to millimeters table, break the given value into its whole number, tenths, hundredths, and thousandths values and convert each separately. Then add the individual values together to form a final equivalence value. For example,

$$3.472 \text{ inches} = \text{how many millimeters?}$$

Whole-number value	$3.000 =$	76.2000	Values
Tenths value	$0.400 =$	10.1600	from
Hundredths value	$0.070 =$	1.7780	Table 13-1
Thousandths value	$0.002 =$	$\underline{0.0508}$	
		88.1888	

Therefore,

$$3.472 \text{ inches} = 88.1888 \text{ mm, or approximately } 88 \text{ mm}$$

Table 13-1 is only good for decimal values. Fractional values must be converted to decimal equivalents before they may be converted into millimeters. For example,

TABLE 13-1 Inches to millimeters.

Whole Numbers		Tenths		Hundreds		Thousands	
in	mm	in	mm	in	mm	in	mm
1	25.4	.1	2.54	.01	.254	.001	.0254
2	50.8	.2	5.08	.02	.508	.002	.0508
3	76.2	.3	7.62	.03	.762	.003	.0762
4	101.6	.4	10.16	.04	1.016	.004	.1016
5	127.0	.5	12.70	.05	1.270	.005	.1270
6	152.4	.6	15.24	.06	1.524	.006	.1524
7	177.8	.7	17.78	.07	1.778	.007	.1778
8	203.2	.8	20.32	.08	2.032	.008	.2032
9	228.6	.9	22.86	.09	2.286	.009	.2286
10	254.0	1.0	25.40	.10	2.540	.010	.2540
11	279.4						
12	304.8						
13	330.2						
14	355.6						
15	381.0						
16	406.4						
17	431.8						
18	457.2						
19	482.6						
20	508.0						
21	533.4						
22	558.8						
23	584.2						
24	609.6						

TABLE 13-2 Millimeters to inches.

mm	in	mm	in	mm	in	mm	in
1	.039	26	1.024	51	2.008	76	2.992
2	.079	27	1.063	52	2.047	77	3.032
3	.118	28	1.102	53	2.087	78	3.071
4	.158	29	1.141	54	2.126	79	3.110
5	.197	30	1.181	55	2.165	80	3.150
6	.236	31	1.221	56	2.205	81	3.189
7	.276	32	1.260	57	2.244	82	3.228
8	.315	33	1.300	58	2.284	83	3.268
9	.354	34	1.339	59	2.323	84	3.307
10	.394	35	1.378	60	2.362	85	3.347
11	.433	36	1.417	61	2.402	86	3.386
12	.472	37	1.457	62	2.441	87	3.425
13	.512	38	1.496	63	2.480	88	3.464
14	.551	39	1.535	64	2.520	89	3.504
15	.591	40	1.575	65	2.559	90	3.543
16	.630	41	1.614	66	2.598	91	3.583
17	.669	42	1.653	67	2.638	92	3.622
18	.709	43	1.693	68	2.677	93	3.661
19	.748	44	1.732	69	2.717	94	3.701
20	.787	45	1.772	70	2.756	95	3.740
21	.827	46	1.811	71	2.795	96	3.780
22	.866	47	1.850	72	2.835	97	3.818
23	.906	48	1.890	73	2.874	98	3.858
24	.945	49	1.929	74	2.913	99	3.898
25	.984	50	1.969	75	2.953	100	3.937

mm	in	mm	in
100	3.937	600	23.622
200	7.874	700	27.559
300	11.811	800	31.496
400	15.748	900	35.433
500	19.685	1000	39.370

$$\frac{3}{8} \text{ inch} = \text{how many millimeters?}$$

$$\frac{3}{8} \text{ inch} = 0.375$$

Whole-number value	0.000 = 0.0000	Values
Tenths value	0.300 = 7.6200	from
Hundredths value	0.070 = 1.7780	Table 13-1
Thousandths value	0.005 = 0.1270	
	9.525	

Therefore,

$$\frac{3}{8} \text{ inch} = 9.525 \text{ mm}$$

To use the millimeters to inches table (Table 13-2), simply look up the value in the table. Fractions of a millimeter are not included. If a fractional millimeter value is required, use the relationship 1 inch = 25.4 mm and calculate the value as shown in Section 13-3. For values greater than 100 mm, look up the hundredths value in the hundredths value table and look up the tenths and units values in the main part of the table; then add the results to form a final equivalent value. For example,

$$537 \text{ mm} = \text{how many inches?}$$

500 mm =	19.685	Values
37 mm =	1.457	from
	21.142	Table 13-2

$$537 \text{ mm} = 21.142 \text{ inches}$$

13-5   FIRST-ANGLE PROJECTIONS

Not only do many foreign countries use a different measuring standard than is used in the United States, they also use a different projection system for presenting orthographic views. The United States uses what is called *third-angle projection*, but many other countries use *first-angle projection*. Figure 13-2 illustrates the differences in the two systems by showing the same object drawn in each. By comparing the two drawings shown in Figure 13-2, we see that the front views in each system are exactly the same. The top views also appear to be the same, although they are located differently relative to the front view. If you are familiar with third-angle projections, you will know that the top view of a first-angle projection appears to be located where the bottom view should be. This apparent reversal of locations comes about because of the way the views are taken. In third-angle projection the viewer looks *at* the object. In first-angle projection the viewer looks *through* the object.

To clarify this concept, study the right-side view of the third-angle projection and the left-end view of the first-angle projection. In the third-angle projection the right-side view is a view taken from the right side of the object, looking into the object, and drawn on the same side of the object as the viewer. In first-angle projection the left-end view is a view taken from the left side of the object, looking through the object, and drawn on the side of the object opposite the viewer. Figures 13-3 and 13-4 are two more examples that compare first- and third-angle projections of the same objects. Study them. *Look into* and *look through* the objects.

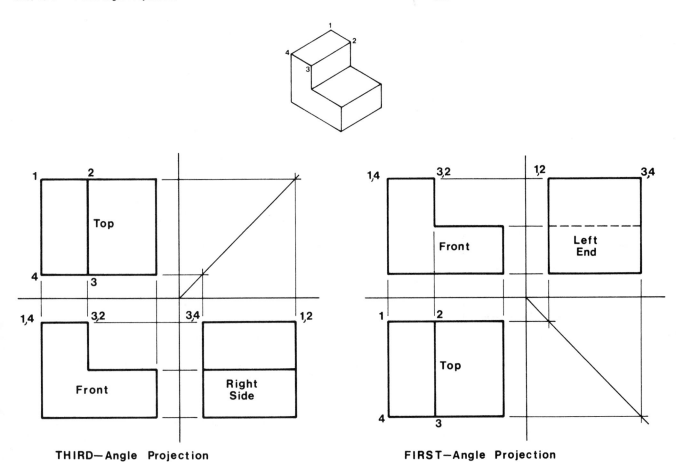

THIRD—Angle Projection

FIRST—Angle Projection

FIGURE 13-2 Comparison between first- and third-angle projections of the same object.

FIGURE 13-3 Comparison between first- and third-angle projections of the same object.

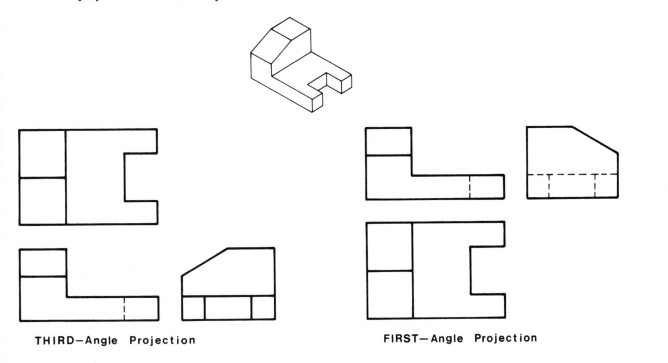

THIRD—Angle Projection

FIRST—Angle Projection

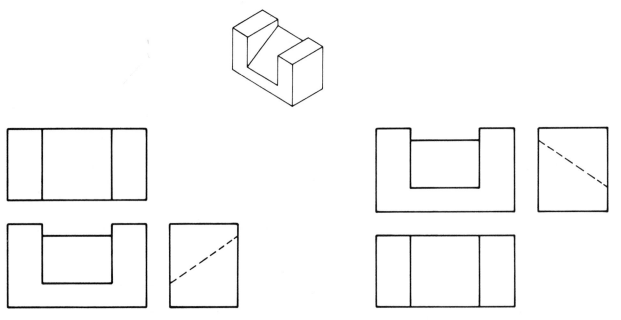

THIRD—Angle Projection   **FIRST—Angle Projection**

FIGURE 13-4 Comparison between first- and third-angle projections of the same object.

From a drawing technique standpoint, the two systems are equally demanding. Visible lines must be heavy and black. Visible lines must be heavier than dimension lines and hidden lines. Lettering must be neat and uniform. The projection theory presented in Chapter 4 is also applicable, although the 45° miter line is located differently (see Figure 13-2).

PROBLEMS

P13-1 Convert the following millimeter values into inches.

(a) 20 mm	(f) 57 mm
(b) 4 mm	(g) 5384 mm
(c) 327 mm	(h) 910 mm
(d) 526 mm	(i) 38 mm
(e) 103 mm	(j) 237 mm

P13-2 Convert the following inch values into millimeters.

(a) 2.378″	(f) 4.125″
(b) 0.750″	(g) 3.500″
(c) 12.875″	(h) 120.000″
(d) 0.020″	(i) 8.820″
(e) 1.006″	(j) 1.324″

P13-3 Convert the following inch values into millimeters:

(a) 1/2″
(b) 2-1/4″
(c) 3-7/8″
(d) 12-5/16″
(e) 5-13/32″

P13-4
through
P13-9
Your company has purchased the rights to produce some parts which up to now have been produced only in Europe. As part of the agreement, the European producer has supplied manufacturing drawings of the parts involved. Convert each drawing, done in millimeters and first-angle projection, into a drawing that can be read by Americans (that is, use decimal inches and third-angle projections).

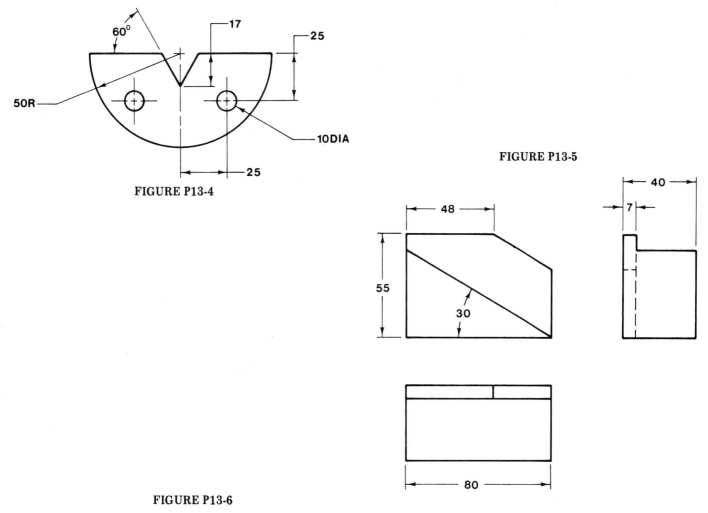

FIGURE P13-4

FIGURE P13-5

FIGURE P13-6

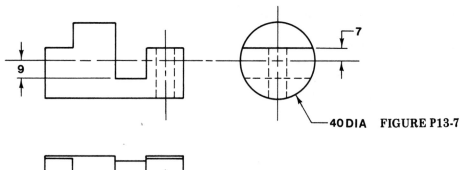

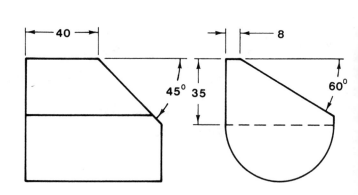

40 DIA   FIGURE P13-7

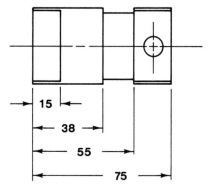

FIGURE P13-8

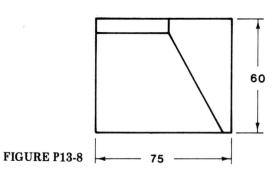

FIGURE P13-9

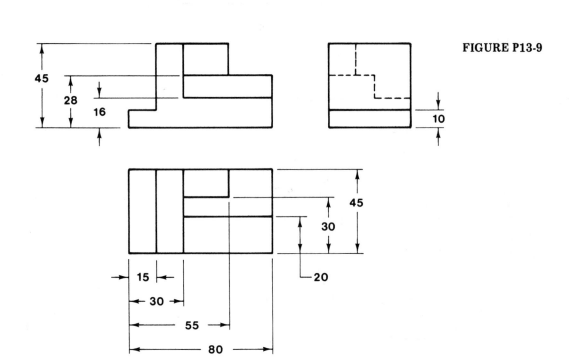

**P13-10
through
P13-13**
Your company has decided to manufacture certain parts in a European plant. To do this, the manufacturing drawings must be converted into the European system of millimeters and first-angle projections. Convert each drawing so that it can be read by Europeans.

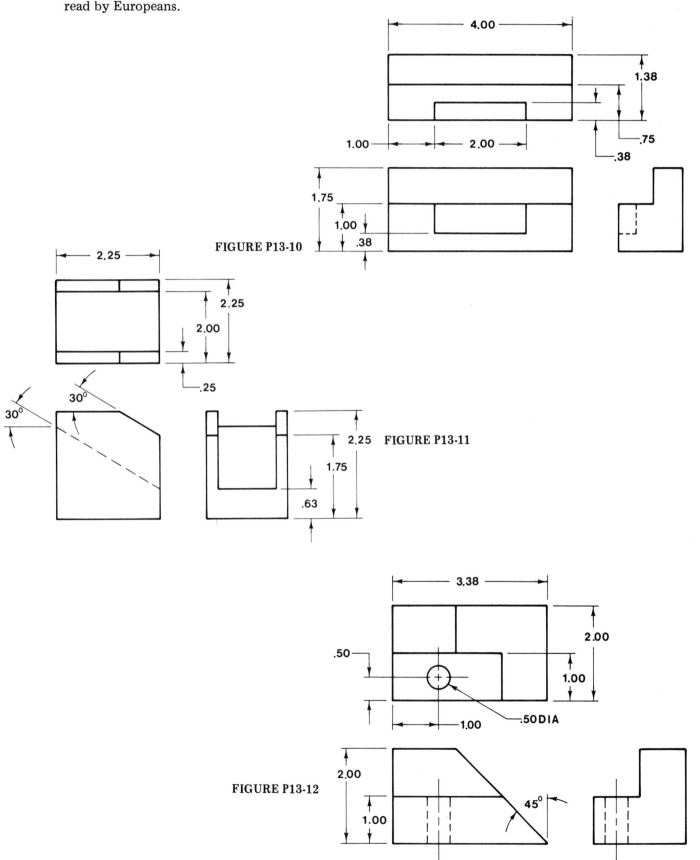

FIGURE P13-10

FIGURE P13-11

FIGURE P13-12

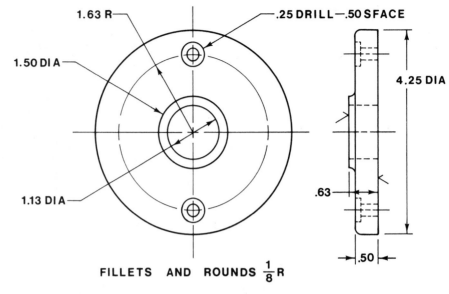

FILLETS   AND   ROUNDS $\frac{1}{8}$ R

FIGURE P13-13

14

PRODUCTION
DRAWINGS

FIGURE 14-0   Courtesy of Teledyne Post, Des Plaines, IL 60016.

14-1 INTRODUCTION

This chapter explains how to prepare production drawings. The term *production drawing* refers to a group of three different types of drawings that are used together to define how to manufacture and assemble objects. The three types of drawings are assembly drawings, detail drawings, and parts lists. Each of these types of drawings has a specific format and drawing rules. The chapter also covers subjects directly related to production drawings, including title blocks, revision blocks, and drawing notes.

14-2 ASSEMBLY DRAWINGS

Assembly drawings are drawings that show the parts of an object in their assembled position, that is, how they fit together. Assembly drawings contain no hidden lines except where absolutely necessary to show how internal parts are assembled. In most cases, sectional views are preferred to hidden lines on assembly drawings.

Figures 14-1 and 14-2 are examples of assembly drawings. Each part is identified by an *item number* enclosed in a circle. Item numbers are referenced to a parts list.

Item 4 in Figure 14-1 is a screw. There are four screws but only one is referenced. If different parts of similar shapes are used on the same assembly drawing, they may be referenced as shown in Figure 14-2 by multiple leader lines. If parts of similar shape are used on the same assembly drawing, but whose size difference makes it easy to tell them apart, only one leader line per item number need be used (see Figure 14-3).

Parts on an assembly drawing should be shown assembled in their natural position (see Figure 14-4). Parts should not be disassembled or in extreme positions as shown, but in a position which clearly shows the object in a natural operating position. If additional clarity is needed, then sectional views, other orthographic views, enlarged details, or notes may be used.

Dimensions may be included on an assembly drawing only when they refer directly to assembly conditions. Similarly, only notes specifically needed for assembly are included.

Assembly drawings are sometimes referred to as *top drawings* because they are the first drawings of a group of production drawings. The procedure is for the assembly drawing to show how the parts fit together and for detail drawings to contain the information needed to manufacture the individual parts. Each part has its own detail drawing, except for standard parts.

The field of an assembly drawing is divided into specific areas for

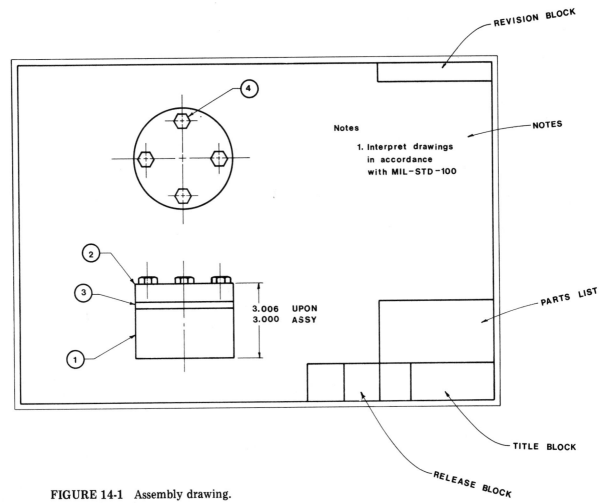

FIGURE 14-1   Assembly drawing.

FIGURE 14-2   Assembly drawing.

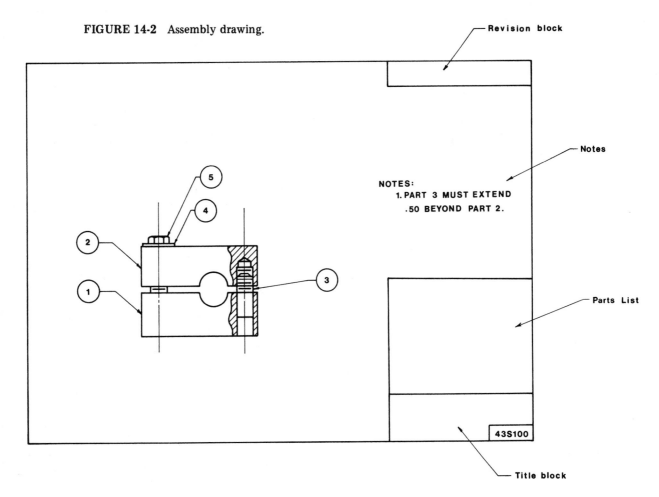

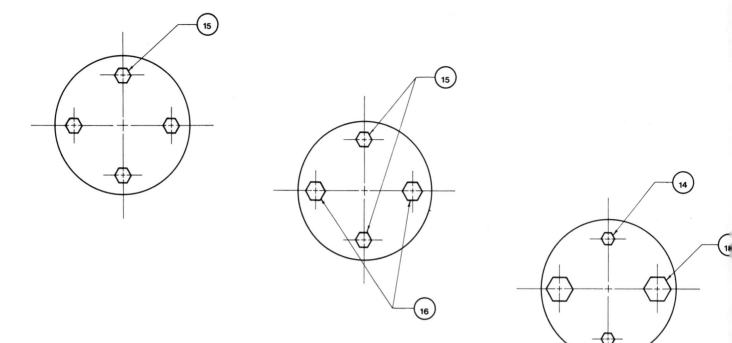

FIGURE 14-3 How to use leader lines with items numbers.

FIGURE 14-4 How to position parts in an assembly drawing.

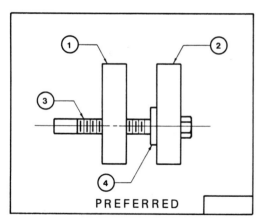

Show Parts in a
Natural Position

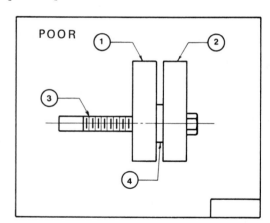

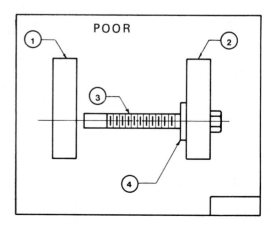

locating the information related to the picture portion of the drawing. The title block is always located in the lower right-hand corner of the drawing as shown in Figure 14-1. The revision block is always in the upper right-hand corner; the parts list, if included on the assembly drawing, is located above the title block, and drawing notes are located above the parts list and below the revision block. If there is no room for drawing notes in this area, they are located just to the left of the parts list. Additional information, such as release blocks, next assembly data, and so on, is located to the left of the title block.

14-3   DETAIL DRAWINGS

Detail drawings are drawings of individual parts and should contain all the information needed to manufacture the part. Figures 14-5, 14-6, and 14-7 are detail drawings of the parts assembled in Figure 14-2.

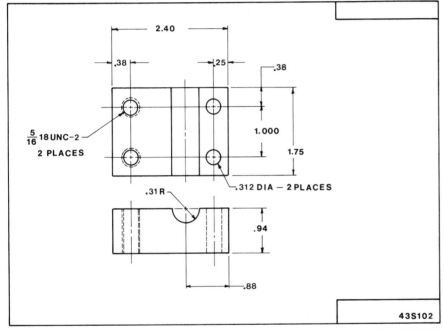

FIGURE 14-5 Sample detail drawing.

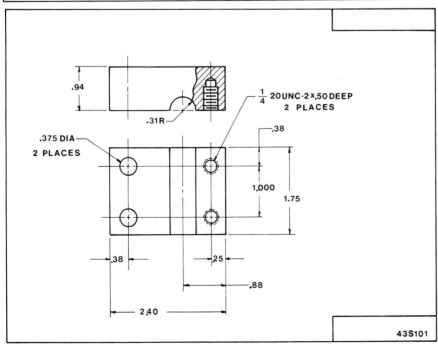

FIGURE 14-6 Sample detail drawing.

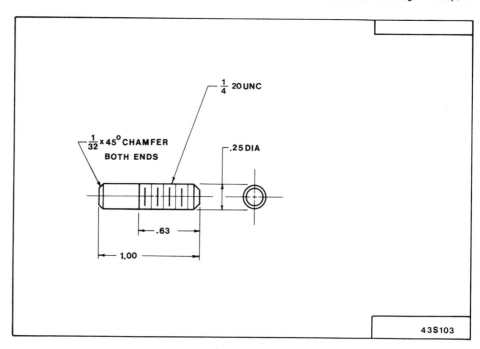

FIGURE 14-7   Sample detail drawing.

Detail drawings contain dimensions, tolerances, material requirements, and any other information needed to make the part. In addition, the detail drawing should include a drawing number which is referenced to the assembly drawing via the parts list.

14-4   PARTS LISTS

A parts list is a listing of all parts needed to assemble an object. Each part is referenced to the assembly drawing by an item number. Figure 14-8 shows a parts list. Figure 14-9 shows the parts list for the assembly shown in Figure 14-2.

In addition to item numbers, a parts list may also contain a part description (name), part number, material specification, note references, and quantity requirements. There is no standard format for parts lists and they vary considerably from company to company.

Figure 14-10 shows the format for the parts list approved for use on government drawings as outlined in DOD-D1000/DOS-STD-100. The 8-inch format is for use on A- and B-size drawings, and the 10.50-inch format is for all other drawing sizes.

Parts lists are prepared from the bottom up. That is, item one is located on the first lower line just above the title block. This is so that new items and revisions can be added to the top of the parts list without interference.

Standard parts, such as screws, nuts, washers, or other purchased parts that are used in an assembly exactly as purchased, without any modification, are listed in the parts list, but do not require detail drawings. For example, item 5 in Figure 14-8 is a hex head screw. This item is considered a standard item because it may be purchased from a manufacturer and used directly in the assembly. It requires no detail draw-

Item No	Nomenclature or Description	Part or ID No	Material/ Specification	Note	Qty
5	.250-20UNC x 1.50 Long Hex Head		SAE 1020 Steel		8
4	.38 x .76 x .06 Washer		SAE 1020 Steel		8
3	Bracket	56574	SAE 1020 Steel	⚠	2
2	Holder	56573	SAE 1020 Steel		1
1	Base	56572	SAE 1020 Steel		1

FIGURE 14-8   A parts list.

NO	DESCRIPTION	MATL	PART NUMBER	QTY
5	SCREW	ST	.31-18 UNC x 1.38 HEX HEAD	2
4	WASHER	ST	.375 x .875 x .083	2
3	GUIDE PIN	SAE 1020	43S103	2
2	TOP CLAMP	SAE 1020	43S102	1
1	BASE CLAMP	SAE 1020	43S101	1

FIGURE 14-9   A sample parts list.

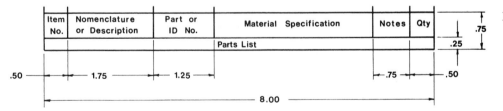

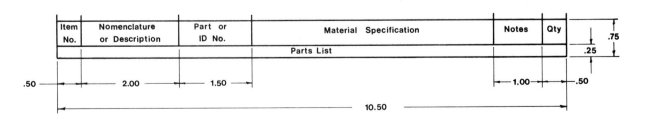

FIGURE 14-10   Parts list formats.

ing. The description of the screw in the parts list must be very exact and include an accurate and complete description of the screw. The standard callout, as defined in Chapter 12, is considered sufficient.

Item 4, a washer, is also considered a standard part and requires only a listing in the parts list. It requires no detail drawing. Washers are identified on a parts list by inside diameter, outside diameter, and thickness.

14-5   TITLE BLOCKS

Title blocks include information necessary to identify the drawing and the company that owns the drawing. Figure 14-11 shows a sample title block. Figure 14-12 shows two title block formats.

Figure 14-13 shows two additional title block formats with two release blocks located adjacent and to the left of the title blocks. Release blocks are used in industry to ensure an orderly checking and documentation procedure for the drawing. As a drawing moves from one department to another, it is "signed off" (initialed) in the release block. When all required signatures have been collected, the drawing is considered ready for production and is released.

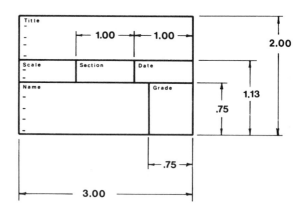

FIGURE 14-11   Sample title block.

FIGURE 14-12   One format for title blocks.

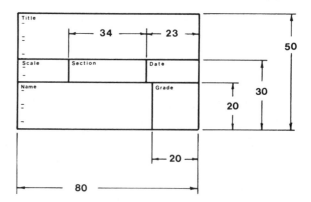

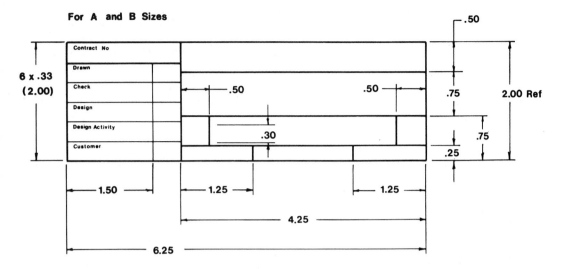

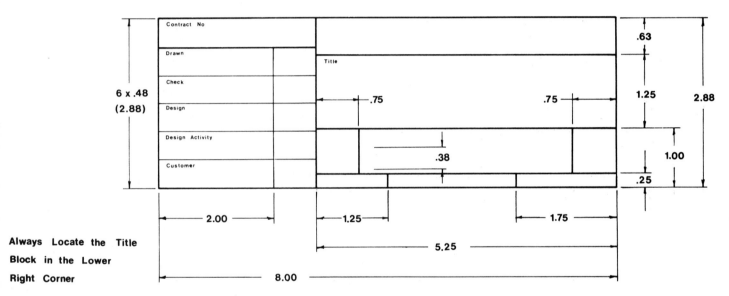

FIGURE 14-13   Two formats for title blocks.

14-6 REVISION BLOCKS

Revision blocks are used to record any changes made to a drawing. The changes may be corrections or simply reflect new design requirements. Figure 14-14 shows a sample revision block. Figure 14-15 shows two formats for revision blocks. Revisions listed in the revision block are cross-referenced to the field of drawing by letters enclosed in triangles (see Figure 14-14).

Revisions are identified in the revision block by using letters in alphabetical order. The letters are also located on the field of the drawing as close as possible to the change. The revision letter is enclosed in a triangle to prevent confusion and to call attention to revision. In Figure 14-14, revison A changed a dimension from 2.00 to 2.13. The letter A, enclosed in a triangle, is located next to the new dimension.

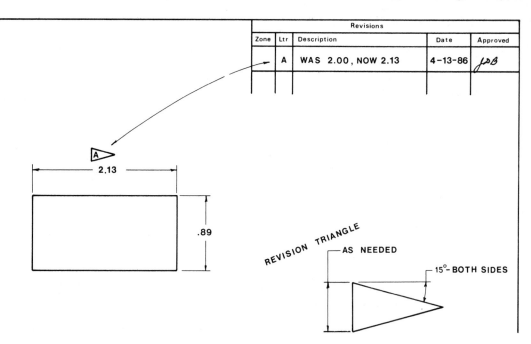

FIGURE 14-14   Sample revision block.

FIGURE 14-15   Two formats for revision blocks.

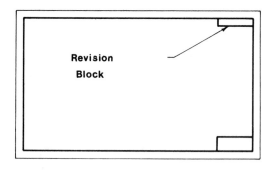

For A and B sizes

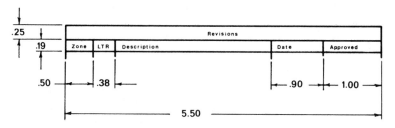

For other than A and B sizes

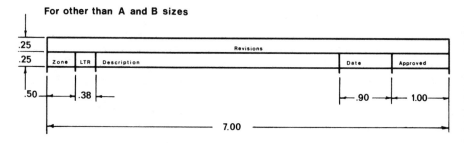

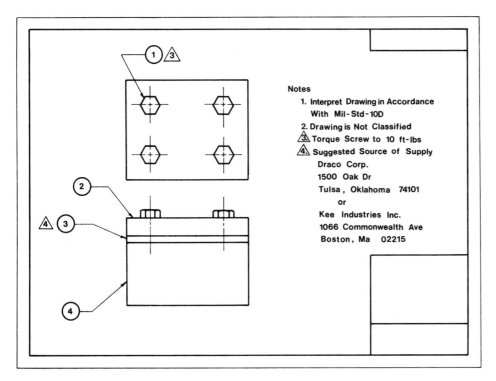

Notes
1. Interpret Drawing in Accordance
   With Mil-Std-10D
2. Drawing is Not Classified
3. Torque Screw to 10 ft-lbs
4. Suggested Source of Supply
   Draco Corp.
   1500 Oak Dr
   Tulsa, Oklahoma   74101
        or
   Kee Industries Inc.
   1066 Commonwealth Ave
   Boston, Ma   02215

FIGURE 14-16   An assembly drawing with drawing notes.

14-7   DRAWING NOTES

Drawing notes are a listing of information needed for correct manufacture of the parts shown on the drawing, but which are not pictorial.
They may include references to appropriate company or government
specifications, or, as shown in Figure 14-16, references to suggested
vendors or assembly information.

Drawing notes are listed by number, in consecutive order. They
are located on the right side of the drawing, above the title block and
below the revision block.

Drawing notes are cross-referenced to the field of the drawing by
locating the note number, enclosed in an equilateral triangle, next to
the appropriate part of the drawing. For example, note 3 specifies
torque requirement for the four screws. The number 3 is enclosed in the
triangle next to item 1, the screw.

14-8   DRAWING ZONES

Large drawings are divided into zones similar to those used on a map.
Letters are used to define the horizontal zones and numbers are used to
define the vertical zones. Figure 14-17 illustrates a zoned drawing.

Zone numbers are usually written in boxes with the letter over the
number as follows:

C/4 , D/2 , A/13 , etc.

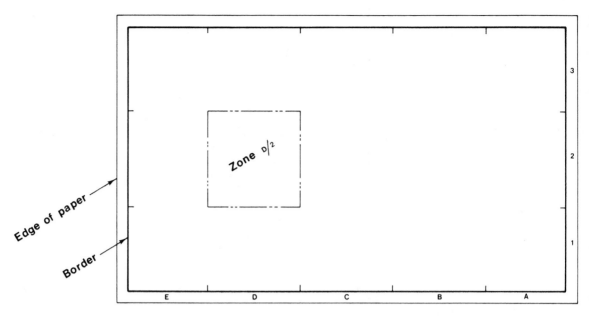

FIGURE 14-17   Example of a zoned drawing.

14-9   ONE-, TWO-, AND PARTIAL-VIEW DRAWINGS

Up to this point we have shown three views of every object. Three views are not always necessary for complete definition of an object and in some cases two views are sufficient. Occasionally, just one view is enough. Figure 14-18 is an example of a two-view drawing. Figure 14-19 is an example of a one-view drawing. In both figures the objects are completely defined and require no other orthographic views.

Unfortunately, there is no rule to follow in determining the number of views needed. Each object must be judged separately according to its individual drawing requirements.

For some objects, one orthographic view and part of another are sufficient for complete definition. A view that includes only part of an

FIGURE 14-18   Example of a two-view drawing. In this case, the side view adds nothing to the drawing and can be eliminated.

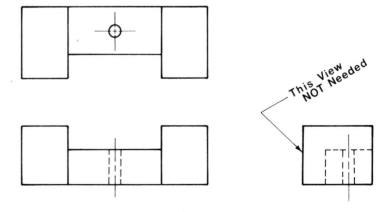

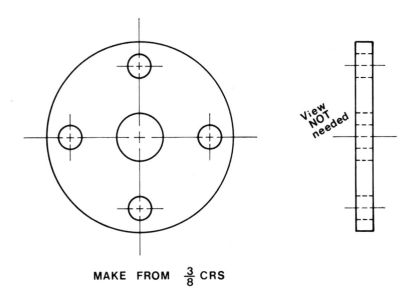

MAKE FROM $\frac{3}{8}$ CRS

FIGURE 14-19 Example of where one view is sufficient to
define the object.

orthographic view is called a *partial view*. When and where to use partial
views is up to the discretion of the drafter, as long as the final drawing
completely defines the object. Figure 14-20 is an example of a drawing
that includes partial views.

To show where a partial view has been broken off (the rest of the
view has been omitted), use a break line. Two kinds of break lines are
used — one for general use and one when break lines are very long.
Figure 14-21 illustrates the two break lines. Figure 14-22 presents an
example of how the long break line is used, and Figure 14-20 illustrates
the general break line. General break lines are drawn freehand as shown
in Figure 14-21.

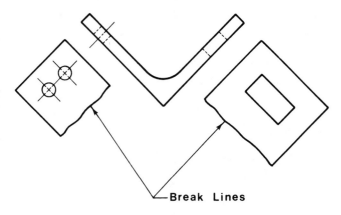

Break Lines

FIGURE 14-20 Example of a drawing that includes partial
views.

FIGURE 14-21 How to draw break lines. The wavy
line used for shorter breaks is drawn freehand, whereas
the line for longer breaks is drawn as shown.

BREAK   LINES

for Shorter Breaks

(thick)

for Long Breaks

(thin)

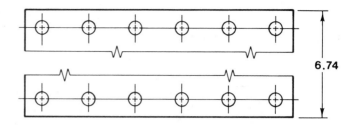

FIGURE 14-22   Example that includes a long break line.

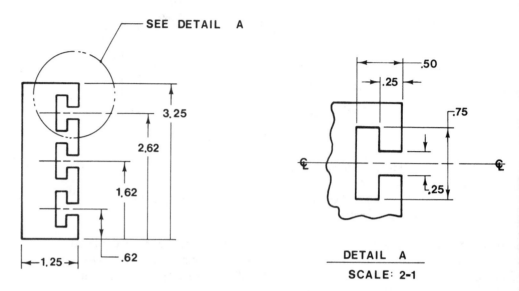

FIGURE 14-23   Example of a drawing detail.

14-10   A DRAWING DETAIL

A drawing detail is a special kind of partial drawing. It is used to enlarge a specific part of a drawing that is too small or too complicated to be completely understood if only shown in its existing size. Figure 14-23 is an example of a drawing that includes a drawing detail.

When you draw a drawing detail, always clearly state the scale used and always label both the detail and the original source of the detail. As with one-view, two-view, and other partial drawings, there is no rule on when a drawing detail should be used. It is up to the drafter to judge his or her drawing and to determine whether or not a drawing detail will help clarify any particular area.

14-11   DRAWING SCALES

Drawing scales are used because some objects are too big to fit on a sheet of drawing paper and others are so small that they could not be seen on a drawing. House drawings, for example, are drawn at a reduced scale. Electronic microcircuits are drawn at an increased scale.

Figure 14-24 shows one full-sized and two scaled drawings of the same square. Note that the scale used is clearly defined.

The scale note 1/2 = 1 means that every 1/2 inch on the drawing is actually 1 inch on the object. In other words, the drawing is one-half the size of the true object size. Similarly, the scale note 2 = 1 means that 2 inches equal 1 inch; thus, the drawing is twice as large as the actual object. The note 1 = 1 means that the drawing is exactly the same size as the object.

When you dimension scaled drawings, never change the stated

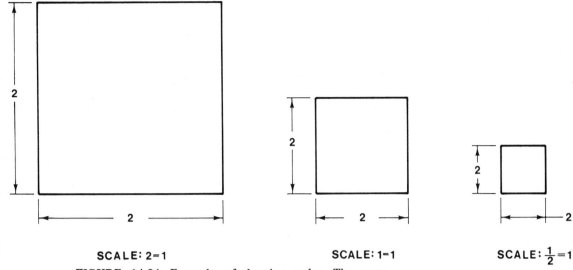

SCALE: 2=1   SCALE: 1=1   SCALE: $\frac{1}{2}$=1

FIGURE 14-24 Example of drawing scales. The same
square has been drawn using three different scales.

dimensions. Only change the size of the picture portion of the drawing.
Look again at Figure 14-24. Note that the object has the same dimensions in each scale despite the change in the drawn size of the object.

PROBLEMS

P14-1 Given the pivot assembly shown in Figure P14-1, prepare an
assembly drawing, a parts list, and any appropriate detail
drawings. If large paper is used, include the parts list on the
assembly drawing. If small paper is used, use a separate sheet
for the parts list. All dimensions are in millimeters.

FIGURE P14-1

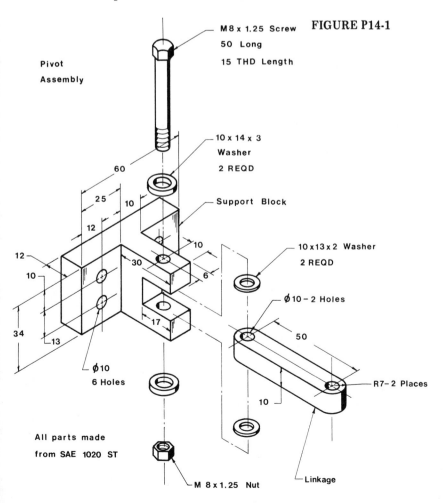

P14-2   Given the guide bracket shown in Figure P14-2, prepare an assembly drawing, a parts list, and any appropriate detail drawings. If large paper is used, include the parts list on the assembly drawing. If small paper is used, use a separate sheet for the parts list. All dimensions are in millimeters.

FIGURE P14-2

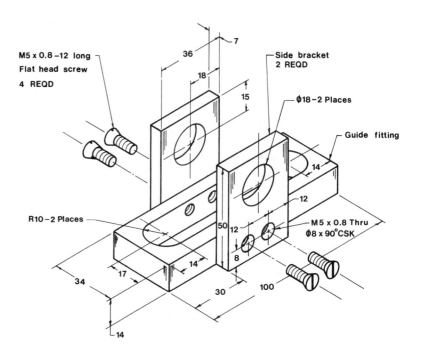

P14-3   Given the pressure chamber shown in Figure P14-3, prepare
an assembly drawing, a parts list, and any appropriate detail
drawings. If large paper is used, include the parts list on the
assembly drawing. If small paper is used, use a separate sheet
for the parts list. All dimensions are in millimeters.

FIGURE P14-3

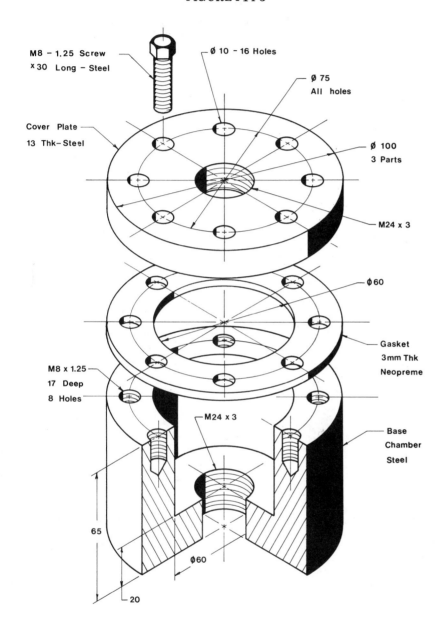

P14-4   Given the round holder shown in Figure P14-4, prepare an assembly drawing, a parts list, and any appropriate detail drawings. If large paper is used, include the parts list on the assembly drawing. If small paper is used, use a separate sheet for the parts list. All dimensions are in inches. The assembly drawing should be prepared on at least 11 × 17 paper. Size all fasteners to match their mating holes. Calculate fastener length from the given hold depths and material thickness.

FIGURE P14-4

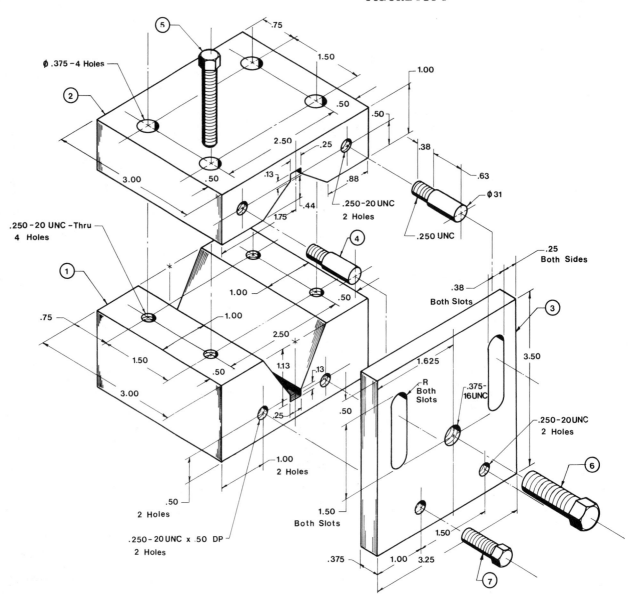

P14-5   Redraw the assembly shown in Figure P14-5. Make any changes that you feel will help clarify the drawing. Also draw detailed drawings of each of the component pieces of the assembly (including the screws). Each square on the grid is 0.20 per side.

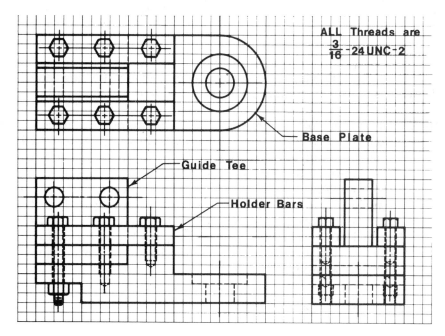

FIGURE P14-5

P14-6 Given the assembly drawing shown in Figure P14-6, prepare detail drawings of each component part. Specify thread sizes for each screw. Assume that there are six screws between the base plate and body and six screws between the cover casting and body. Each square on the grid pattern is 1/8 per side.

FIGURE P14-6

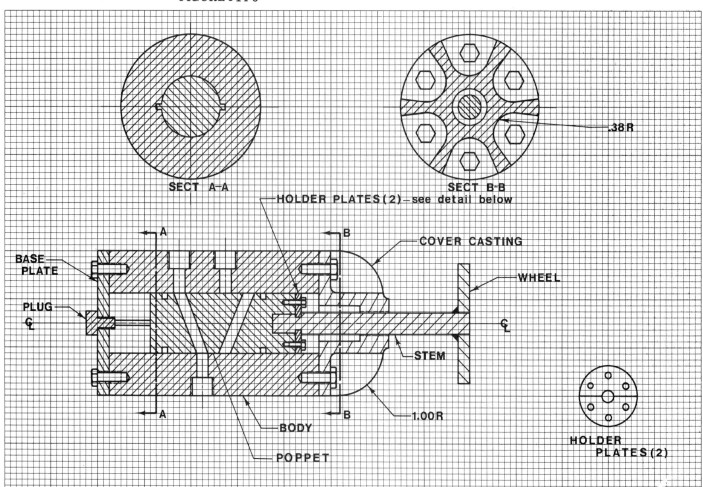

DESIGN PROBLEMS

For each of the following designs, prepare an assembly drawing and detail drawings of each component part, and a parts list.

P14-7   Given the exploded drawing of a holding fixture and details of each of the fixture's component parts shown in Figure P14-7, draw a complete assembly drawing of the fixture. Use whatever views (orthographic, sectional, and so on) are necessary for complete definition of how all the pieces are to be assembled. Add a note to the assembly drawing to have the latch pin, dowel pins, cam pivot pin, and cover pivot pin peined after assembly. Also add a parts list to the assembly drawing which includes a complete listing of parts required for the assembly.

FIGURE P14-7(a)

Latch

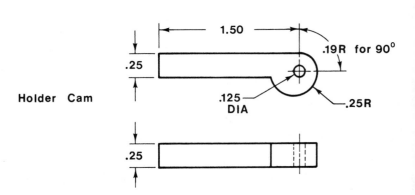

Holder Cam

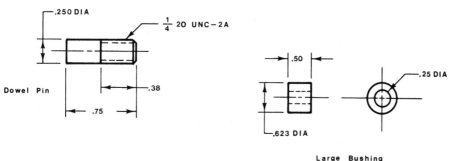

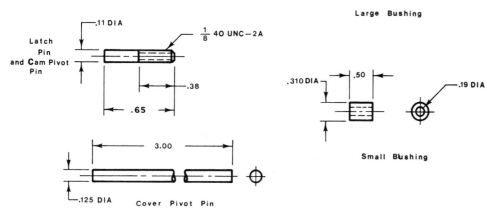

Dowel Pin

.250 DIA

¼ 20 UNC–2A

.38

.75

Latch Pin and Cam Pivot Pin

.11 DIA

⅛ 40 UNC–2A

.38

.65

3.00

.125 DIA   Cover Pivot Pin

Large Bushing

.50

.623 DIA

.25 DIA

Small Bushing

.310 DIA

.50

.19 DIA

FIGURE P14-7(b)

FIGURE P14-7(c)

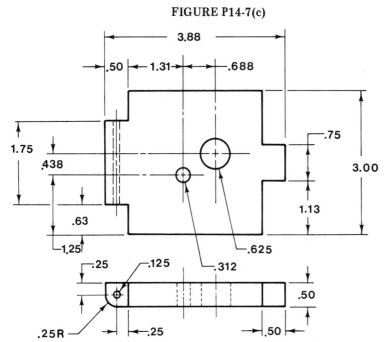

3.88

.50   1.31   .688

1.75

.438

.63

1.25

.75

3.00

1.13

.625

.312

.25   .125   .25R

.25   .50

.50

Cover Plate

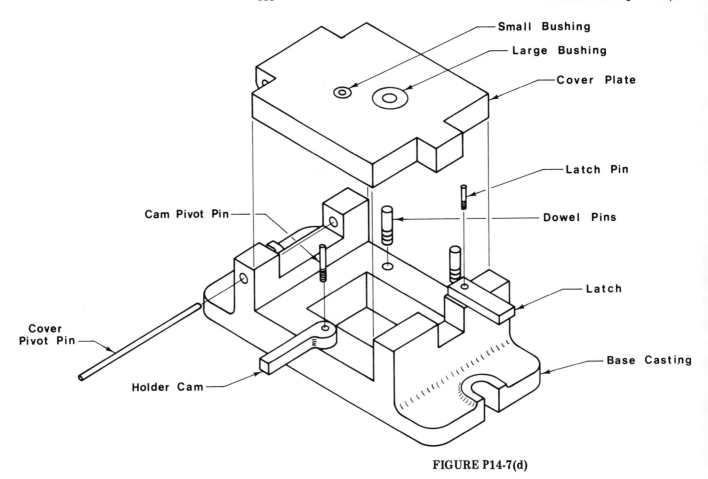

Small Bushing

Large Bushing

Cover Plate

Latch Pin

Cam Pivot Pin

Dowel Pins

Latch

Cover
Pivot Pin

Base Casting

Holder Cam

FIGURE P14-7(d)

FIGURE P14-7(e)

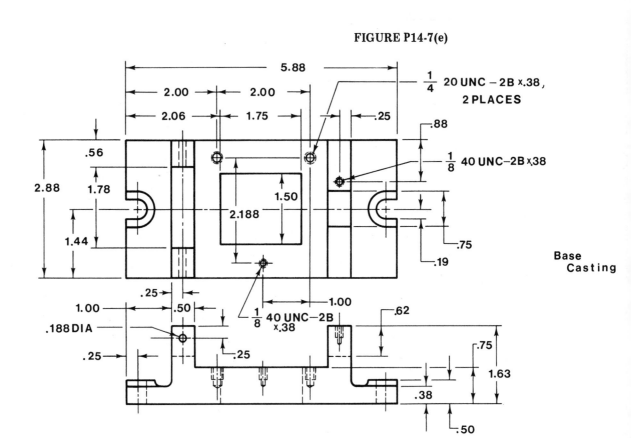

5.88

2.00   2.00

2.06   1.75   .25

$\frac{1}{4}$ 20 UNC – 2B x.38,
2 PLACES

.88

.56

$\frac{1}{8}$ 40 UNC–2B x.38

2.88   1.78

1.50

2.188

1.44

.75

.19

Base
Casting

.25

1.00   .50

1.00

.62

.188 DIA

$\frac{1}{8}$ 40 UNC–2B
x.38

.75

1.63

.25   .25

.38

.50

P14-8   Design a four-shelf bookcase.

P14-9   Design a drawing table.

P14-10   Design a case for carrying all your drafting equipment (do not include drafting machines).

P14-11   Prepare a detail drawing of any standard tool (hammer, wrench, etc.).

P14-12   Prepare an assembly drawing and detail drawings of each component part of a ball-point pen or leadholder.

P14-13   Design a wine rack. Allow for bottles of at least three different sizes.

P14-14   Design a portable, removable food tray for use while eating in a car. Specify the make of car for which you are designing the tray.

15

ISOMETRIC
DRAWINGS

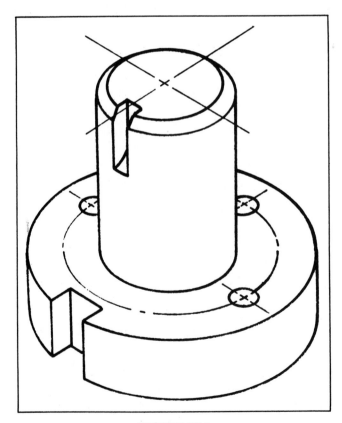

FIGURE 15-0

15-1 INTRODUCTION

Isometric drawings are technical pictures that can be drawn by using instruments. They are not esthetically perfect pictures because their axes do not taper as they approach infinity. Figure 15-1 shows a comparison between an isometric drawing of a rectangular box and a pictorial drawing (such as an artist would draw) of the same object, and it demonstrates the distortion inherent in isometric drawings. Note how the back corner of the isometric appears much larger than the back of the pictorial drawing. Despite this slight distortion, isometric drawings are a valuable way to convey technical information.

The basic reference system for isometric drawings is shown in Figure 15-2. The three lines are 120° apart and may be thought of as a vertical line and two lines 30° to the horizontal, which means that they may be drawn by using a 30–60–90 triangle supported by a T-square. All isometric drawings are based on this axis system.

Normally, an isometric drawing is positioned so that the front, top, and right-side views appear as shown in Figure 15-3. This may be varied according to the position that the drafter feels best shows the object.

FIGURE 15-1 Comparison between an isometric drawing and a pictorial drawing. Note the visual distortion of the top rear corner of the isometric drawing.

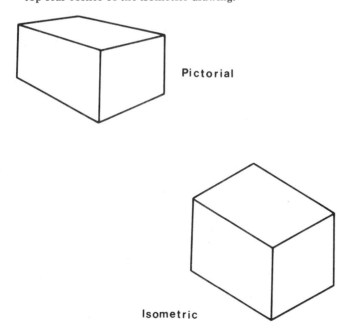

Pictorial

Isometric

An Isometric Axis

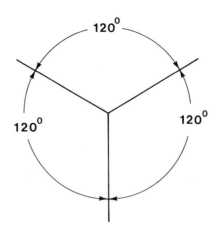

Can also be drawn

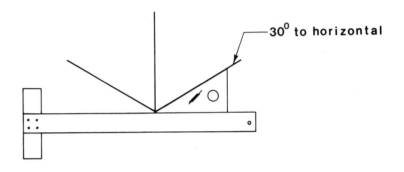

An Isometric Cube

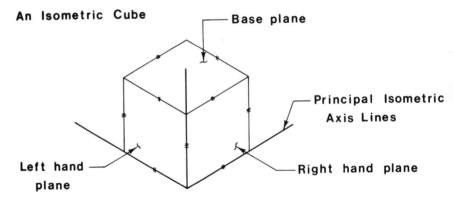

Lines marked //, ///, ○ are parallel

FIGURE 15-2 Basic reference system for isometric drawings.

Dimensional values are transferable from orthographic views only to the axis, or lines parallel to the axis, of isometric drawings. Angles and inclined dimensional values are not directly transferable and require special supplementary layouts which are explained in this chapter.

Isometric drawings do not normally include hidden lines, although hidden lines may be drawn if special emphasis of a hidden surface is required.

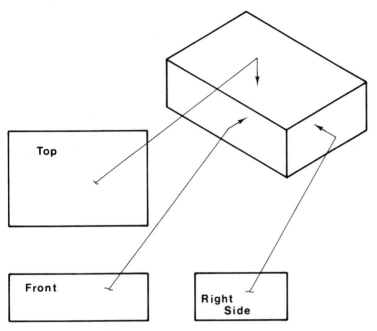

FIGURE 15-3 Definition of the relationship between the front, top, and side views as drawn orthographically and isometrically.

15-2   NORMAL SURFACES

Figure 15-4 is a sample problem that requires you to create an isometric drawing from given orthographic views. Since all surfaces in the problem are normal (90° to each other), all dimensional values may be transferred directly from the orthographic views to the isometric axis, or lines parallel to the isometric axis. The basic height, width, and length of the object are 1-1/2, 2, and 3, respectively, in both the isometric and orthographic drawings. Figure 15-5 is the solution to Figure 15-4 and was derived by the following procedure:

GIVEN: Front, top, and side views.
PROBLEM: Draw an isometric drawing.

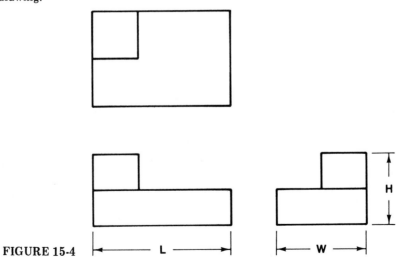

FIGURE 15-4

SOLUTION:

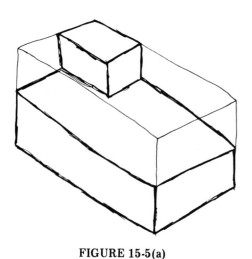

FIGURE 15-5(a)

1. Make, to the best of your ability, a freehand sketch of the solution. See Section 5-7 for instructions on how to make sketches. Remember that since it is easier to make corrections and changes on a sketch than on a drawing, you should make your sketch as complete and accurate as possible.

2. Using very light lines, lay out a rectangular box whose height, width, and length correspond to the height, width, and length given in the orthographic views.

3. Using very light lines, lay out the specific shape of the object. Transfer dimensional values directly from the orthographic views to the axis, or lines parallel to the axis, of the isometric drawing.

4. Erase all excess lines and smudges; carefully check your work; and then darken in all final lines to their proper color and pattern.

FIGURE 15-5(b)

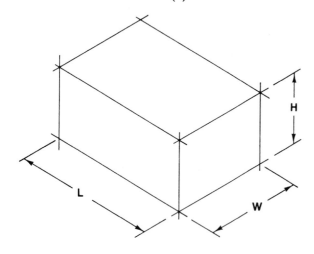

FIGURE 15-5(c)

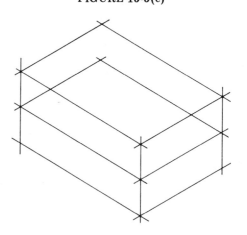

FIGURE 15-5(d)

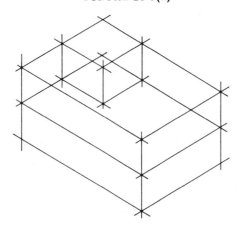

FIGURE 15-5(e)

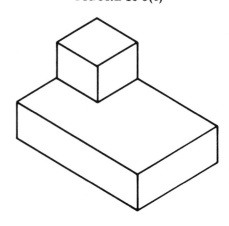

Figure 15-6 is another example of an isometric drawing created from given orthographic views and including only normal surfaces. Figure 15-7 is the solution to Figure 15-6.

GIVEN: Front, top, and side views.
PROBLEM: Draw an isometric drawing.   SOLUTION:

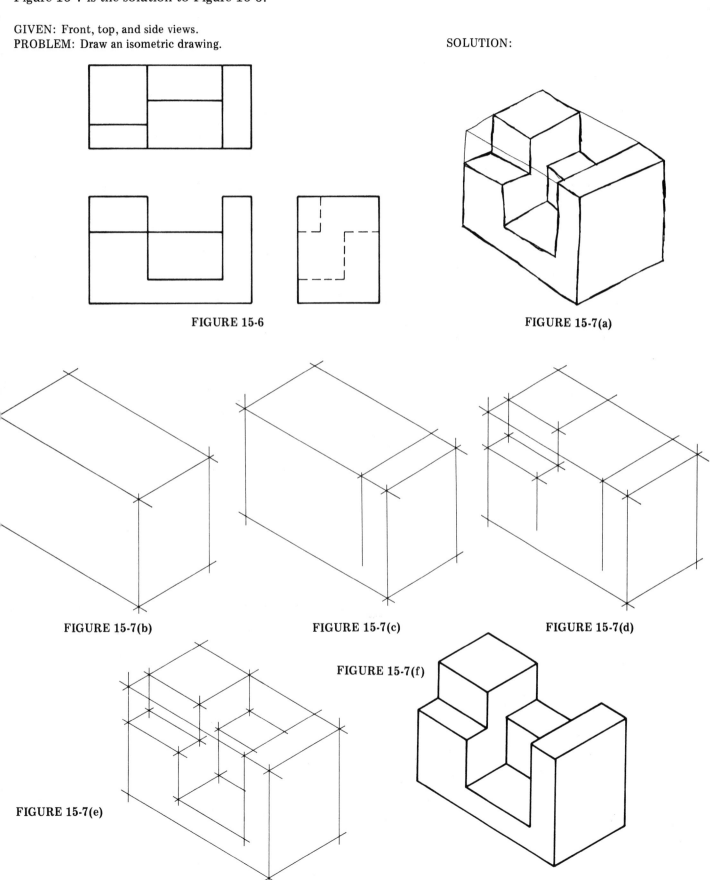

FIGURE 15-6   FIGURE 15-7(a)

FIGURE 15-7(b)   FIGURE 15-7(c)   FIGURE 15-7(d)

FIGURE 15-7(f)

FIGURE 15-7(e)

15-3 SLANTED AND OBLIQUE SURFACES

Figure 15-8 is a sample problem that involves the creation of an isometric drawing from given orthographic views that contain a slanted surface. The slanted surface is dimensioned by using an angular dimension. Angular dimensions cannot be directly transferred from orthographic views to isometric drawings.

To transfer an angular dimensional view from an orthographic view to an isometric drawing, convert the angular dimensional value to its component linear value and transfer the component values directly to the axis of the isometric drawing. Figure 15-9 illustrates this procedure by showing two angular dimensions that have been converted to their respective component linear values and then showing how these values are transferred to the isometric axis. Normally, a drafter simply measures the full-sized orthographic views and then transfers the information, but if this information is not available, he or she makes a supplementary layout from which the necessary values may be measured. Supplementary layouts may be made on any extra available paper and should be saved for reference during the checking of the drawing.

Angles cannot be measured on an isometric drawing using a protractor. The resulting angle will be wrong, as shown in Figure 15-10. Special isometric protractors are available, but they are sometimes difficult to use. The best procedure is, as shown in Figure 15-11, to create a freehand sketch, convert the angular components to linear components using a supplementary layout, then draw the isometric drawing.

GIVEN: Front, top, and side views.
PROBLEM: Draw an isometric drawing.

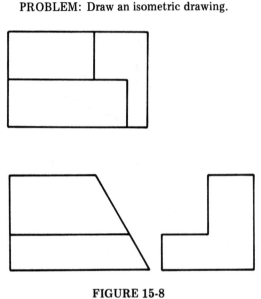

FIGURE 15-8

FIGURE 15-9 Two examples of angular dimensions that have been redimensioned using their linear coordinates. The linear coordinates have been transferred to an isometric axis.

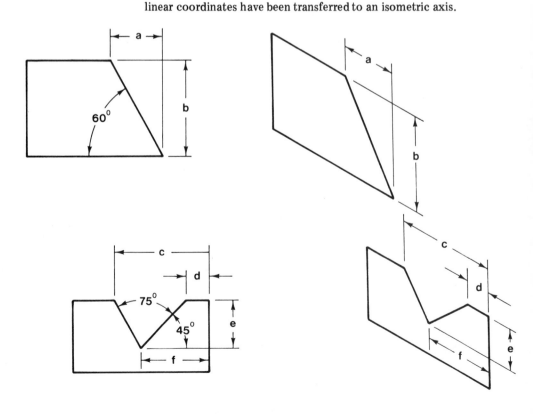

Angular dimension

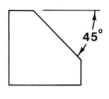

Linear dimensions

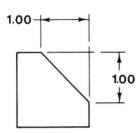

Angular dimensions cannot be transferred
to an isometric drawing using a protractor.
Only linear dimensions can be transferred.

WRONG

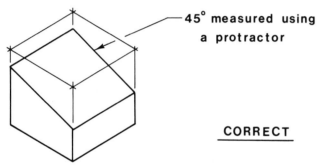

45° measured using
a protractor

CORRECT

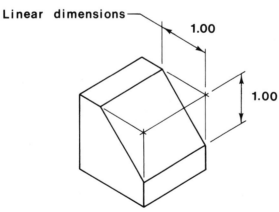

Linear dimensions

FIGURE 15-10 Circular protractors cannot be used on
isometric drawings.

Orthographic Views

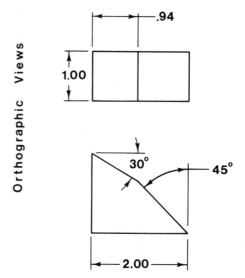

Supplementary Layout

Freehand Sketch

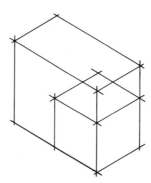

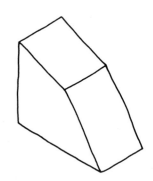

Layout

Final Drawing

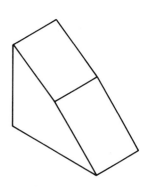

FIGURE 15-11 Change angular dimensions to linear dimensions, then transfer the linear values to the isometric drawing.

Figure 15-12 is the solution to Figure 15-8 and was derived by the following procedure:

SOLUTION:

1. Make, to the best of your ability, a freehand sketch of the solution.
2. Using very light lines, lay out a rectangular box whose height, width, and length correspond to the height, width, and length given in the orthographic views.

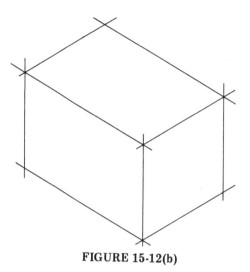

FIGURE 15-12(b)

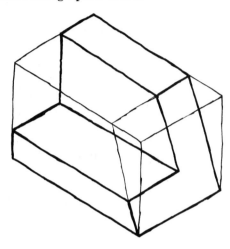

FIGURE 15-12(a)

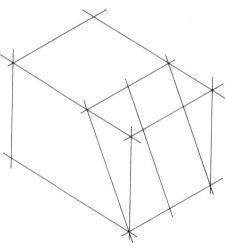

FIGURE 15-12(c)

3. Using very light lines, lay out the specific details of the object. Where necessary, make supplementary layouts that furnish the linear component values which you can transfer to the isometric axis. In this case, the 30° component layout is shown in Figure 15-9.
4. Erase all excess lines and smudges; check your work; and then draw in all lines to their proper color and pattern.

FIGURE 15-12(d)

FIGURE 15-12(e)

FIGURE 15-12(f)

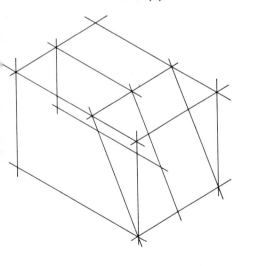

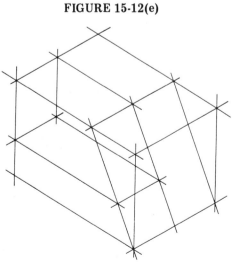

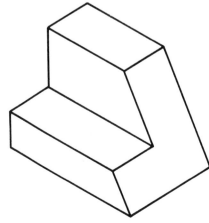

Figure 15-13 is a sample problem that requires you to make an isometric drawing from given orthographic views that include an oblique surface. The solution was derived by using basically the same procedure that was used for slanted surfaces. As with angular dimensional values, the dimensional values that define an oblique surface must be converted to their respective linear component values before they may be transferred to the isometric axis. If necessary, supplementary layouts should be made. Figure 15-14 is the solution to Figure 15-13.

GIVEN: Front, top, and side views.
PROBLEM: Draw an isometric drawing.

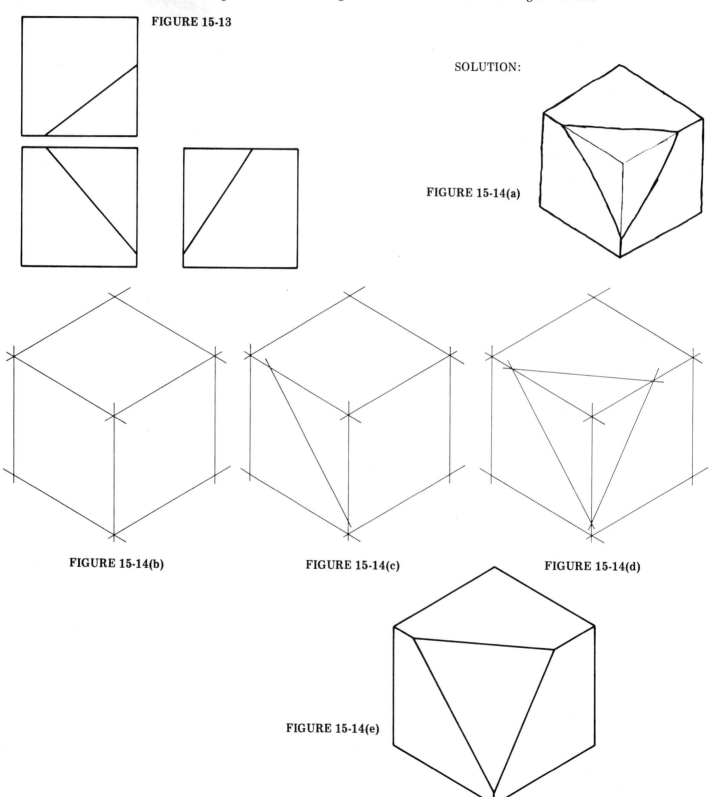

FIGURE 15-13

SOLUTION:

FIGURE 15-14(a)

FIGURE 15-14(b)

FIGURE 15-14(c)

FIGURE 15-14(d)

FIGURE 15-14(e)

15-4   HOLES IN ISOMETRIC DRAWINGS

There are two basic methods for drawing holes for isometric drawings. One method is to use instruments and draw the holes by using the four-center ellipse method. The other method is to use an isometric hole template as a guide. The template is much easier and faster to use, but templates are available only in standard hole sizes. Very large or odd-sized holes may only be drawn by using the four-center ellipse method.

The four-center ellipse method is presented in Figure 15-17(a). When you use this method, be careful that the four centers are located accurately. If the centers are not located properly, the four individual arcs will not meet to form a smooth, continuous ellipse. A good practice that will help you draw a smooth continuous ellipse is to lightly construct the ellipse and then check it for accuracy before drawing in the final heavy arcs.

An isometric hole template may be conveniently used as a guide for drawing the hole size for which it is cut. Figure 15-15(a) illustrates an isometric hole template. To align the template for drawing, first draw in the hole centerlines, and then align the guide lines printed on the template adjacent to the desired hole with the centerlines on the drawing. Figure 15-15(b) shows this procedure.

The outside shape of an isometric hole template is cut to a specific shape to help make it easier to align it. Figure 15-15(b) shows how the template can be positioned along a horizontal straight edge to help

FIGURE 15-15   (a) Isometric hole template; (b) How to position an isometric hole template.

a.

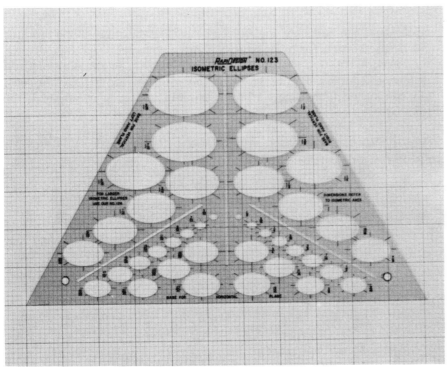

Setup for Horizontal Plane

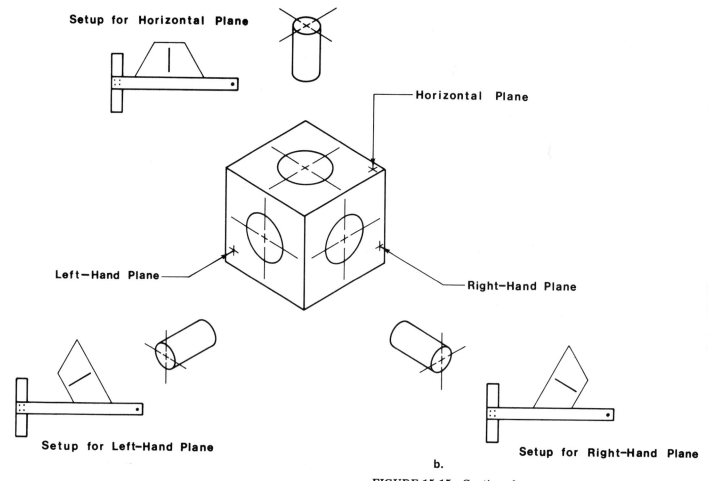

Horizontal Plane

Left—Hand Plane

Right—Hand Plane

Setup for Left—Hand Plane

b.

Setup for Right—Hand Plane

FIGURE 15-15   Continued

ensure that the holes are aligned correctly. The elliptical shape produced can be used either for holes or for cylindrical shapes.

Figure 15-16 is a problem that requires you to draw a hole in an isometric drawing. Figure 15-17 is the solution using the four-center ellipse method, and Figure 15-18 is the solution using an isometric hole template.

GIVEN: Front and side views.
PROBLEM: Draw an isometric drawing.

FIGURE 15-16

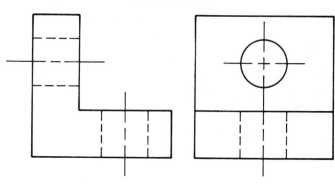

SOLUTION   Using the four-center ellipse method:

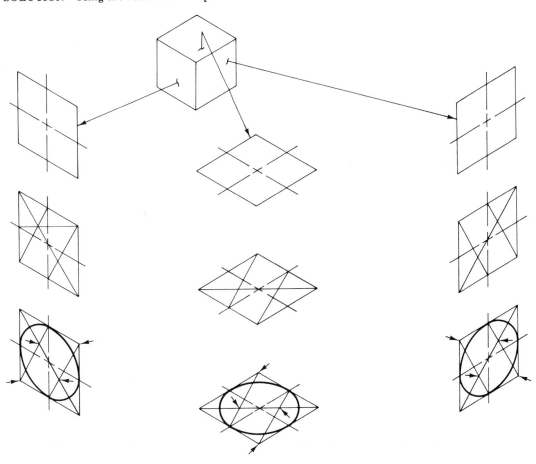

←Indicates the location of compass center points

FIGURE 15-17(a) Four-center method for drawing isometric elipses. *Note:* This method is good only for isometric
drawings; use the approximate elipse method described in
Section 3-26 for all other elipses.

FIGURE 15-17(b)

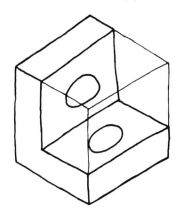

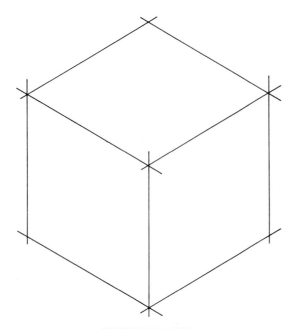

FIGURE 15-17(c)

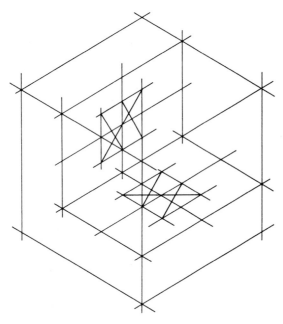

FIGURE 15-17(e)

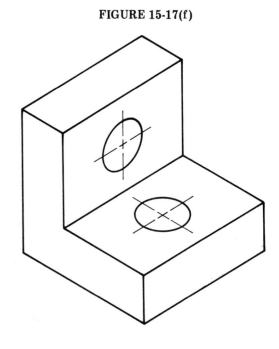

FIGURE 15-17(d)

FIGURE 15-17(f)

SOLUTION   Using an isometric hole template:

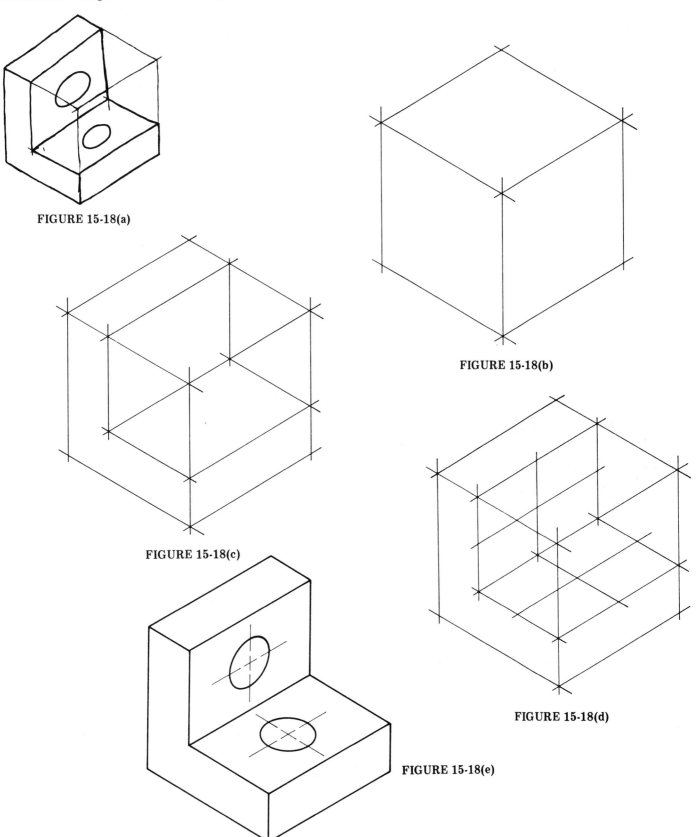

FIGURE 15-18(a)

FIGURE 15-18(b)

FIGURE 15-18(c)

FIGURE 15-18(d)

FIGURE 15-18(e)

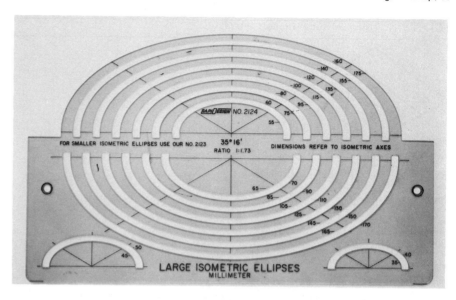

FIGURE 15-19   A template for large isometric ellipses.

Very large or very small holes can be drawn using special templates, although the sizes are limited to those available on the templates. Figure 15-19 shows a template designed to help draw large isometric ellipses. Only half an ellipse can be drawn at a time, so the template must be set up in two opposite and matching positions as shown in Figure 15-20. A 30-60-90 triangle may be used to help align the template when drawing right- and left-plane ellipses.

Very small ellipses can be drawn with the help of the template shown in Figure 15-21. The template is aligned as shown in Figure 15-22. A 30-60-90 triangle is used to help align the template in the right and left planes.

BASE PLANE

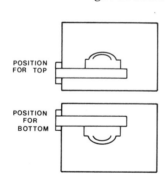

FIGURE 15-20   How to position the template shown in Figure 15-21.

LEFT PLANE

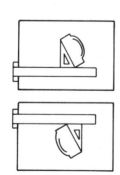

RIGHT PLANE

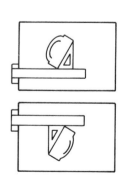

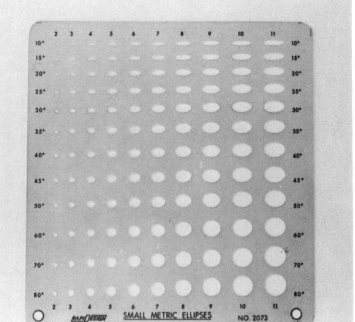

FIGURE 15-21   A template for very small ellipses.

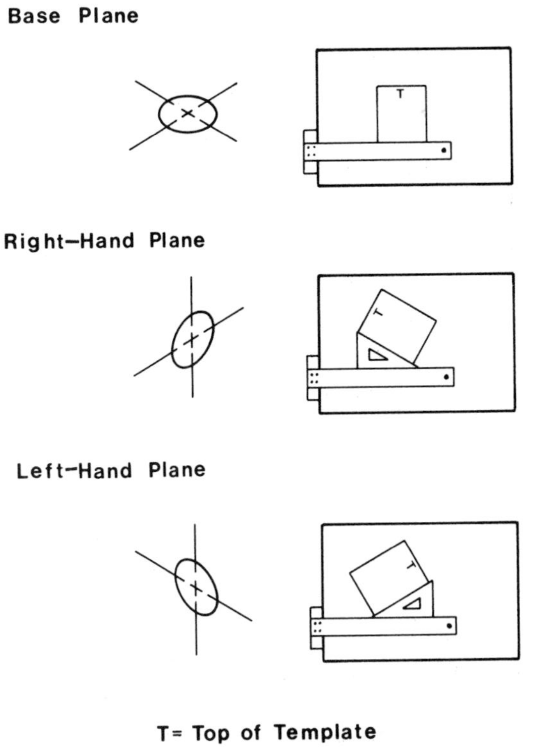

Base Plane

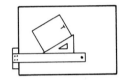

Right—Hand Plane

Left-Hand Plane

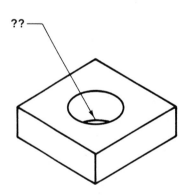

T= Top of Template

FIGURE 15-22 How to position the template shown in Figure 15-21.

In drawing a hole for an isometric drawing there arises the question of whether or not the bottom edge of the hole can be seen. If it can be seen, how much of it can be seen? Figure 15-23 illustrates the problem.

To determine exactly if and how much of the bottom edge of the hole should be drawn, locate the center point of the hole on the bottom surface and draw in the hole by using the same procedure you used for the hole on the top surface. If the hole drawn on the bottom surface appears within the hole on the top surface, it should appear on the finished drawing. If the hole drawn on the bottom surface does not appear within the hole on the top surface, it should not appear on the finished drawing. Figure 15-24 presents a sample problem that illustrates this procedure. Figure 15-25 is the solution to Figure 15-24.

GIVEN: A front view.
PROBLEM: Draw an isometric drawing.

??—

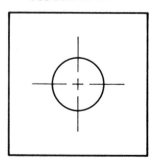

FIGURE 15-23 When does the bottom edge of a hole show in an isometric drawing?

FIGURE 15-24

SOLUTION:

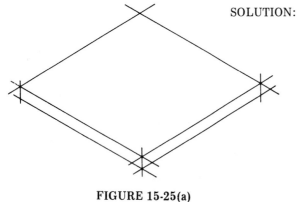

FIGURE 15-25(a)

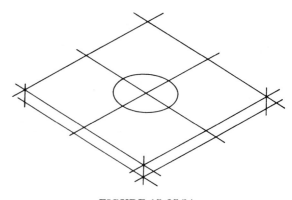

FIGURE 15-25(b)

FIGURE 15-25(c)

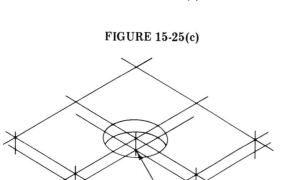

Centerpoint for
Bottom Surface

FIGURE 15-25(d)

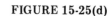
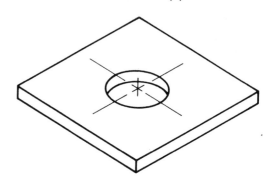

15-5 ROUND AND IRREGULAR SURFACES

Figure 15-26 is a sample problem that requires you to create an isometric drawing from given orthographic views that contain a round surface. To make an isometric drawing of a round surface, use either an isometric template for a guide or the point method as described in this section. Figure 15-27 is a solution to Figure 15-26 that was derived by using an isometric ellipse template. Figure 15-28 is a solution that was derived by using the point method. The procedures are as follows.

GIVEN: Front, top, and side views.
PROBLEM: Draw an isometric drawing.

FIGURE 15-26

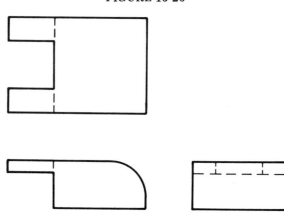

SOLUTION   Using an isometric ellipse template:

1. Define on one of the orthographic views (the one that shows the round surface as part of a circle) the center point of the round surface and the intersections of the centerlines with the surfaces of the object. In this example the center point is marked 0, and the two intersections are marked points 1 and 2.

2. Draw a rectangular box and transfer the points 1, 2, and 0 to the front plane of the isometric drawing and label them 3, 4, and 5.

3. Project the points in the front plane across the isometric drawing to the back plane.

4. Align the proper hole in the isometric ellipse template with the centerlines on the front isometric surface, and draw in the isometric arc. Repeat the same procedure for the back surface.

5. Erase all excess lines and smudges; check your work; draw in the remaining lines of the object lightly at first and then darken them to their proper color and pattern.

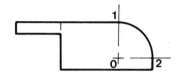

FIGURE 15-27(a)

FIGURE 15-27(b)

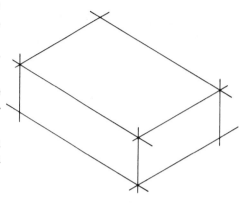

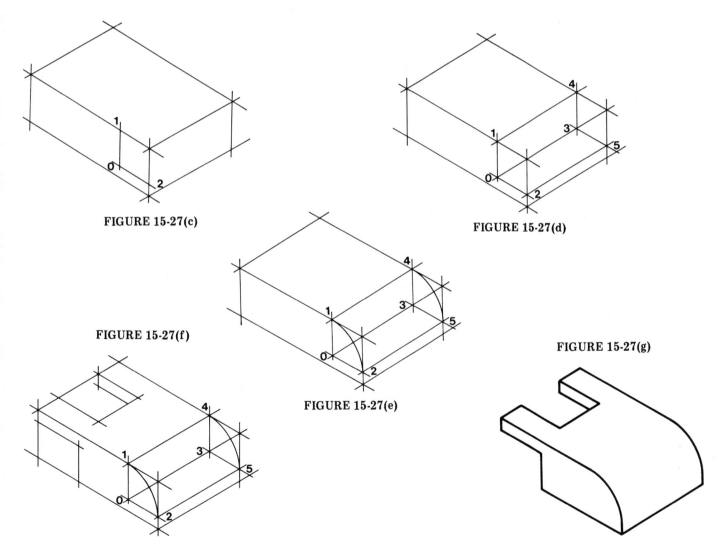

FIGURE 15-27(c)

FIGURE 15-27(d)

FIGURE 15-27(f)

FIGURE 15-27(e)

FIGURE 15-27(g)

SOLUTION   Using the point method:

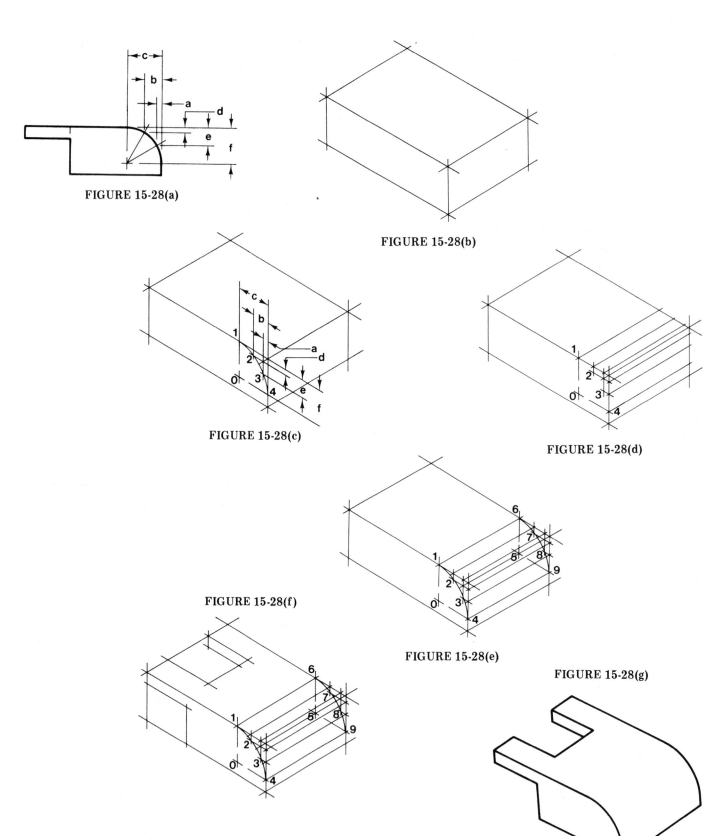

FIGURE 15-28(a)

FIGURE 15-28(b)

FIGURE 15-28(c)

FIGURE 15-28(d)

FIGURE 15-28(f)

FIGURE 15-28(e)

FIGURE 15-28(g)

1. On one of the orthographic views (the one that shows the round surface as part of a circle) mark off a series of points along the rounded surface. The points need not be equidistant. The more points you take, the more accurate will be the final isometric ellipse. If necessary, make a full-sized supplementary layout.

2. Dimension each point horizontally and vertically as shown.

3. Transfer the dimensional values to the isometric axis as shown.

4. Using a French curve as a guide, draw in the isometric arc.

5. Transfer the points to the back of the surface, and, again using a French curve as a guide, draw in the isometric arc.

6. Erase all excess lines and smudges; check your work; draw in the remaining lines of the object lightly at first and then darken them to their final color and pattern.

Figure 15-29 is a sample problem that requires you to draw an isometric drawing from given orthographic views that contain an irregular surface. The point method described for drawing isometric drawings of round surfaces is directly applicable to the creation of isometric drawings of irregular surfaces provided that we use two of the orthographic views to locate the points. Two views are required because the surface may not be parallel to any of the principal planes. Figure 15-30 is the solution to Figure 15-29.

GIVEN: An object.
PROBLEM: Draw an isometric drawing of the object.

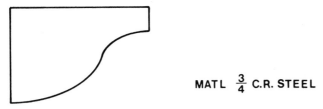

MATL $\frac{3}{4}$ C.R. STEEL

FIGURE 15-29

SOLUTION:

FIGURE 15-30(a)

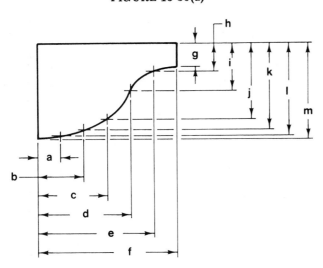

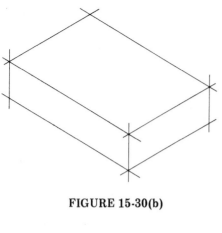

FIGURE 15-30(b)

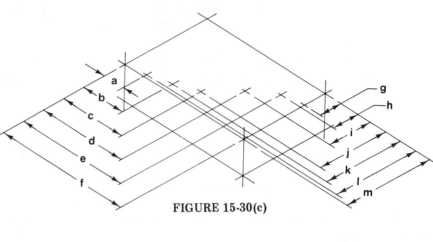

FIGURE 15-30(c)

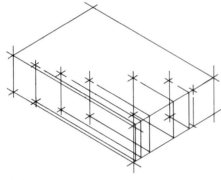

FIGURE 15-30(d)

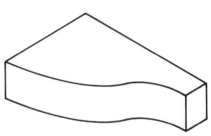

FIGURE 15-30(e)

15-6   FASTENERS

Threaded fasteners are drawn in isometric drawings using a series of isometric ellipses as shown in Figure 15-31. The shading shown is optional. Threads are represented by an isometric ellipse. The first and last threads are drawn using a thinner line, about half as thick as the object lines. Drawing callout notes are interpreted as presented in Chapter 12 and shown in Figure 15-32.

Fasteners that have hexagon-shaped heads, such as screws, bolts, and washers, can be drawn with the help of an isometric hex head and nuts template as shown in Figure 15-33. The template is cut in elliptical shapes, half-hexagon shapes, and a third shape referred to as an umbrella shape.

The umbrella shapes cut into the template can be oriented into any one of the six possible isometric axis orientations as shown in Figure 15-34. In each orientation, the first step is to draw an appropriate isometric ellipse. The ellipse is then used to align the template.

Figure 15-35 shows how to draw a hex head fastener in the right isometric plane. Both an isometric ellipse and hex head template were used. The drawing was created as follows:

1. Locate the centerpoint for the hex head.
2. Draw an isometric ellipse in the correct orientation of diameter equal to the distance across the flats of the fastener.

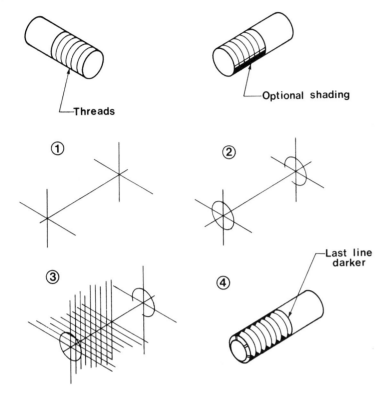

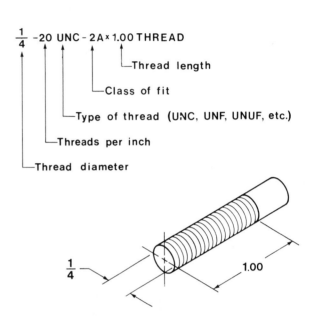

FIGURE 15-31   How to draw threads in isometric.

FIGURE 15-32   How to interpret a thread callout.

FIGURE 15-33   A template for drawing hexagon-shaped fasteners in isometric.

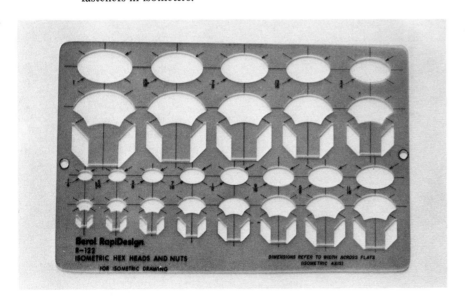

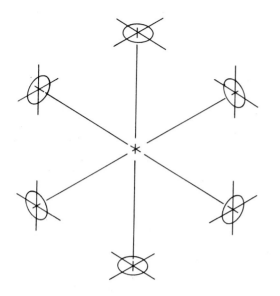

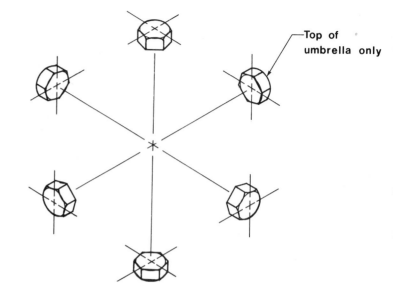

FIGURE 15-34 How to use the template shown in Figure 15-33.

FIGURE 15-35 How to draw a hex head fastener in the right isometric plane.

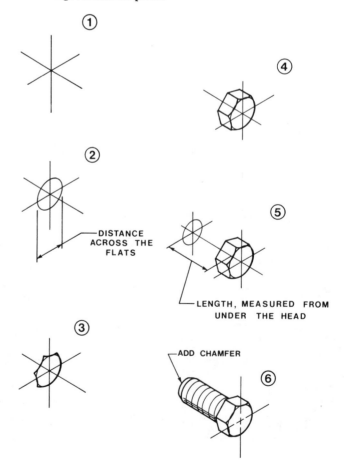

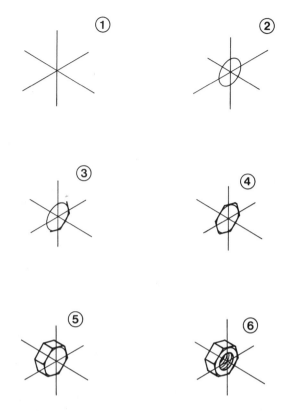

FIGURE 15-36   How to draw a hex nut in isometric.

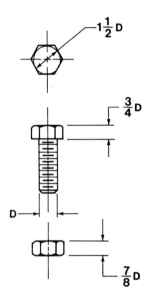

FIGURE 15-37   The general proportions for a hex screw and nut.

3. Align the half-umbrella shape with the elliptical shape and draw in the umbrella shape around the elliptical shape.

4. Align the half-hexagon shapes with the points on the umbrella shape and draw in as shown. The appropriate half-hexagon shape is located directly under the umbrella shape on the template.

5. Measure and mark the required fastener length. Remember that the length of a fastener is measured from under the head.

6. Add the threads and chamfer as shown. A chamfer may be drawn by using a smaller ellipse than was used for the threads. It can be located by eye. The shading shown is optional.

Figure 15-36 shows the procedure for drawing a nut in the right isometric plane. The first step is to set up an appropriate isometric ellipse and then circumscribe it using the half-hex shape found on the template.

Approximate fastener sizes are determined using the general values shown in Figure 15-37. In all the values shown, the D is the nominal diameter of the fastener. Figure 15-38 shows an application of the approximate values shown in Figure 15-37.

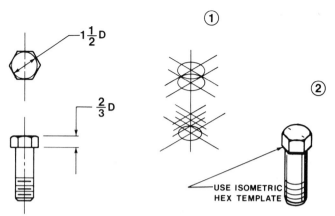

FIGURE 15-38 How to apply the general proportions shown in Figure 15-37.

15-7 WASHERS

Washers are drawn on isometric drawings using isometric ellipses as shown in Figure 15-39. To draw each type of washer, the first step is to define the washer location using an isometric axis. The appropriate isometric ellipses are then added.

Washer sizes are defined by their thickness, inside diameter, and outside diameter as shown in Figure 15-40. The shading shown is optional.

FIGURE 15-39   How to draw a washer in isometric.

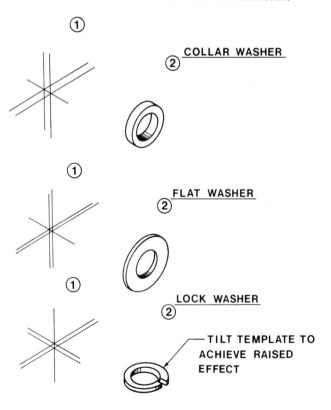

FIGURE 15-40   A sample drawing callout for a washer.

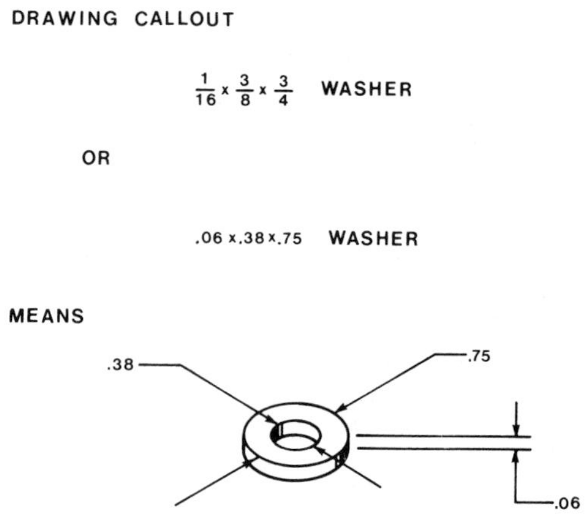

DRAWING CALLOUT

$$\frac{1}{16} \times \frac{3}{8} \times \frac{3}{4} \quad \text{WASHER}$$

OR

.06 x .38 x .75   WASHER

MEANS

15-8   SHADING

Shading and shadows are not usually added to isometric drawing. Shading and shadows are considered part of technical illustration and not technical drawing. This section shows some basic shading and shadowing techniques because drafters are sometimes asked to help prepare technical illustrations for reports or proposals. The techniques shown are very simple and the interested student is referred to books on technical illustration for further information.

Figure 15-41 shows shading on a bolt-and-washer combination. The light source would be to the left. Many other illustrations in this book have been done using the same technique.

Figure 15-42 shows a simple technique for casting a shadow from a cube. The light source is upper left from the cube. A horizontal line is drawn from the lowest corner of the cube to the right. A 45° line is then drawn from the lowest corner of the top surface located directly over the corner used to reference the horizontal line. This is done for all other corners on the right side of the cube, including the unseen back corner as shown. The intersection of the horizontal and 45° lines is then connected, defining the outline of the shadow. The shadow is colored in and the right plane of the cube is crosshatched.

The technique presented in Figure 15-42 represents only one of many different shadow-casting techniques. Many other styles can be created including airbrushing, watercolor washes, and so on.

Figures 15-43 and 15-44 show the technique described above applied to different geometric shapes. Note how the coloring and crosshatching were interchanged in the two illustrations. The figures were drawn using an isometric axis system.

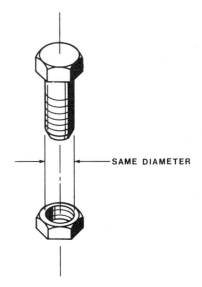

FIGURE 15-41 Sample shading for a bolt and washer.

FIGURE 15-43 How to cast a shadow.

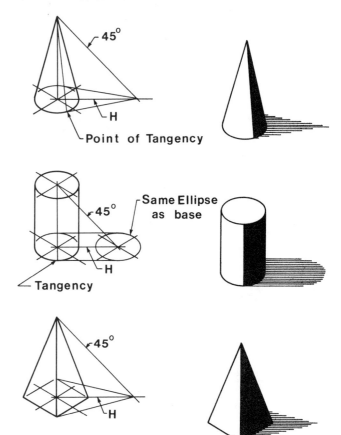

FIGURE 15-42 How to cast a shadow.

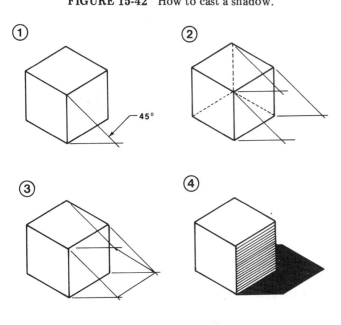

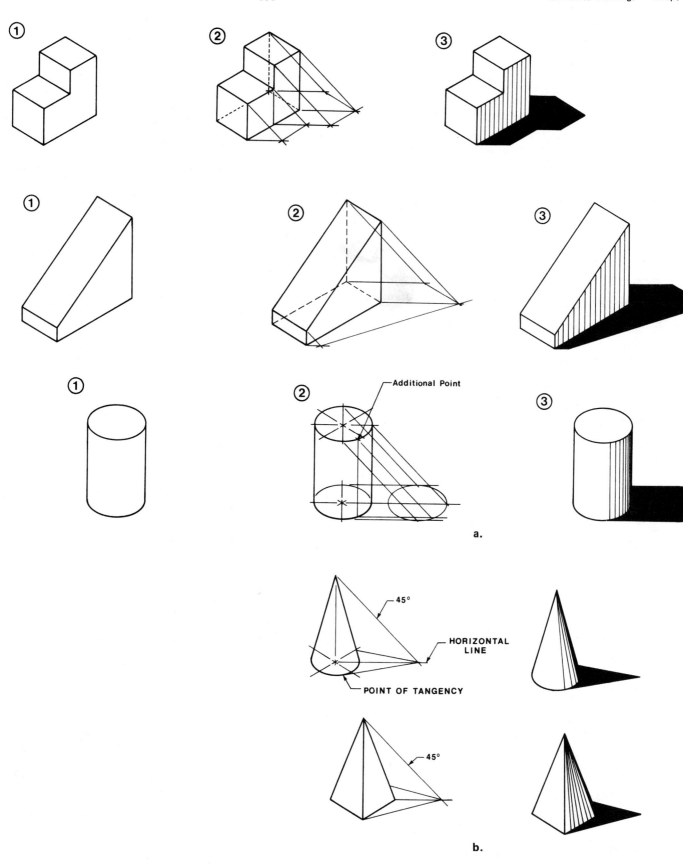

FIGURE 15-44 How to cast a shadow.

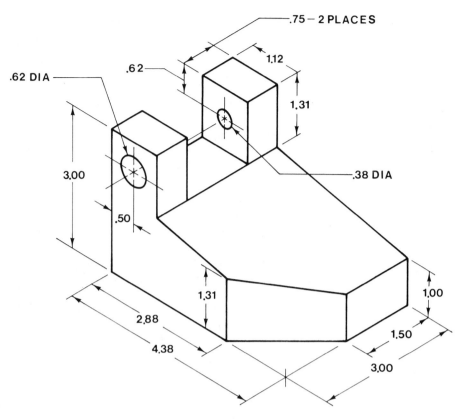

FIGURE 15-45 Example of an isometric drawing dimensioned using the unidirectional system.

15-9   ISOMETRIC DIMENSIONS

Isometric drawings may be dimensioned by using either the aligned system or the unidirectional system. All isometric drawings in this book are dimensioned by using the unidirectional system. Section 6-4 gives a further explanation of the differences between the two systems.

Regardless of the system used, the leader lines must be drawn in the same isometric plane as the surface they are defining. The guide lines for the dimensions in the aligned system are drawn parallel to the edge being defined while the guide lines for the unidirectional system are always horizontal. Figure 15-45 is another example of the unidirectional system. The numbers are drawn 1/8 to 3/16 in both systems.

FIGURE 15-46 Isometric section cut.

15-10   ISOMETRIC SECTIONAL VIEWS

Isometric sectional views are used for the same reasons that orthographic sectional views are used — to clarify objects by exposing important internal surfaces that would otherwise be hidden from direct view. Figure 15-46 shows a full isometric sectional view and a half isometric sectional view. Note that, as with orthographic sectional views, hidden lines are omitted and the crosshatching lines are drawn medium to light in color, 3/32 apart at an inclined angle. Isometric sectional views do not require a defining cutting plane and are usually presented as in-

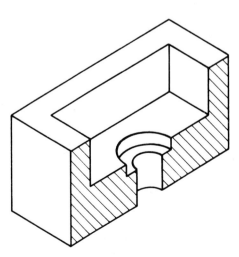

dividual pictures with no accompanying reference drawing. Dimensions are placed on an isometric sectional view in the same way they are for regular isometric drawings.

15-11   AXONOMETRIC DRAWINGS

Isometric drawings are actually just one of a broad category of drawings called *axonometric drawings*. An axonometric drawing is a pictorial drawing, drawn with instruments, that uses some initially defined axis system which remains parallel to infinity.

There are three kinds of axonometric drawings: isometric, dimetric, and trimetric. The classification of an axonometric drawing depends on its axis system. An isometric axis has three equal angles (120°), a dimetric axis has two equal angles, and a trimetric axis has no equal angles. Figure 15-47 shows examples of the three axonometric axes. The oblique drawing, which is covered in Chapter 16, is a special form of trimetric drawing.

An adjustable triangle, such as the one shown in Figure 15-48, is useful when you are creating axonometric drawings because it may be set to any angle, thereby eliminating the need for constant measuring with a protractor.

FIGURE 15-47 Examples of the three different types of axonometric drawings: isometric, dimetric, and trimetric.

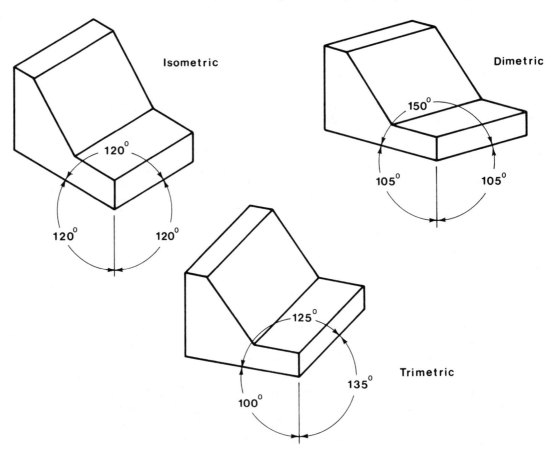

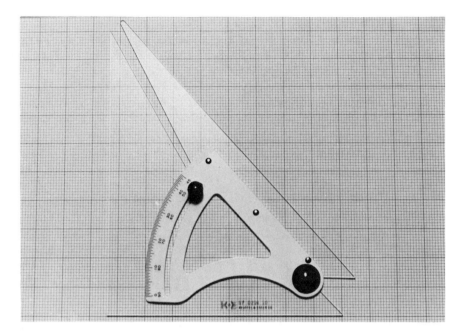

FIGURE 15-48   Adjustable triangle.

15-12   EXPLODED DRAWINGS

Figure 15-49 is an example of an exploded drawing. Exploded drawings are useful because they enable the reader to visualize and understand technical information without requiring him or her to have a knowledge of orthographic projections. They are particularly well suited to assembly drawings because they easily show the relationship between the various parts.

Exploded drawings may be drawn by using any one of the axonometric axis systems provided that the system chosen helps present the information clearly. Exploded drawings rarely contain dimensions or

FIGURE 15-49   Exploded drawing.

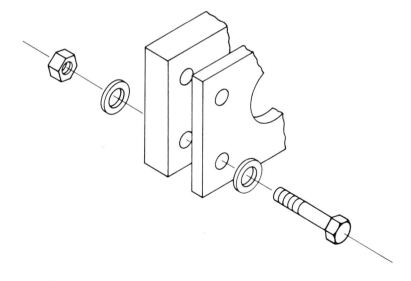

hidden lines because they are usually intended more to be pictures of technical information than actual technical drawings. Parts in an exploded drawing are always labeled either by name or by part number.

Figure 15-50 is an example of an isometric exploded drawing. It was prepared from the assembly drawing of the object. See Figure 15-51, the parts list, and Figures 15-52 and 15-53, the detail drawings.

The bolt, nut, and washer are standard items and have no detail drawings. They are drawn in isometric using the procedures presented in Sections 15-6 and 15-7.

FIGURE 15-50   An exploded drawing.

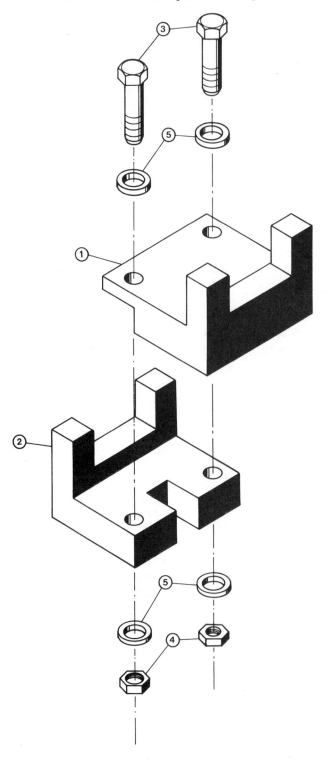

The exploded drawing (Figure 15-50) is really a series of isometric drawings of each of the individual parts of the assembly. They have been spaced so that no part covers another and each part has been identified using the same item numbers listed in the parts list. The shading is as defined in Section 15-8 and is optional.

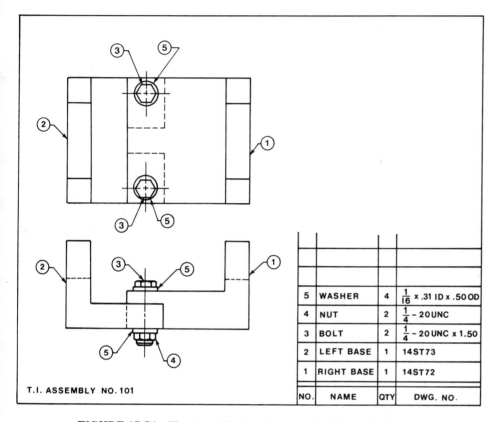

5	WASHER	4	$\frac{1}{16}$ x .31 ID x .50 OD
4	NUT	2	$\frac{1}{4}$ – 20 UNC
3	BOLT	2	$\frac{1}{4}$ – 20 UNC x 1.50
2	LEFT BASE	1	14ST73
1	RIGHT BASE	1	14ST72
NO.	NAME	QTY	DWG. NO.

T.I. ASSEMBLY NO. 101

FIGURE 15-51   The assemble drawing used for Figure 15-50.

FIGURE 15-52   A detail drawing used for Figure 15-50.

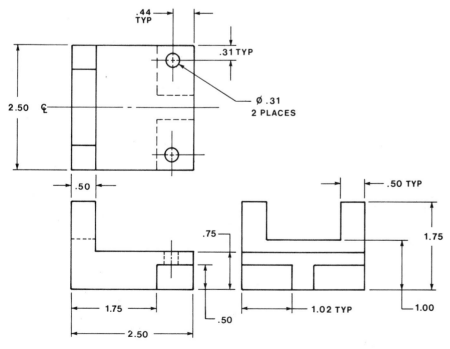

RIGHT BASE
DWG. NO. 14ST72

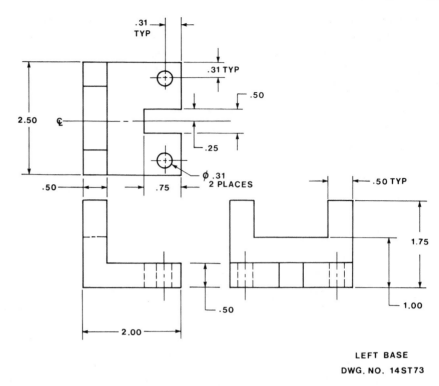

LEFT BASE
DWG. NO. 14ST73

FIGURE 15-53   A detail drawing used for Figure 15-50.

PROBLEMS

P15-1   Create isometric drawings of each object. Dimetric and tri-
through   metric drawings may also be created if assigned by your in-
P15-15   structor.

FIGURE P15-1

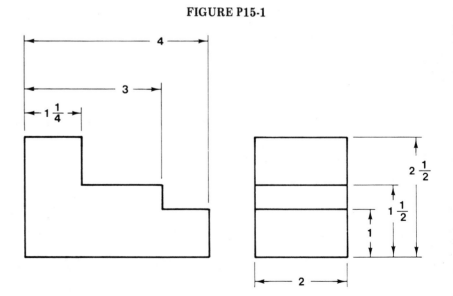

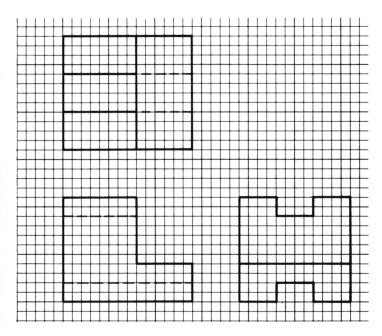

FIGURE P15-2

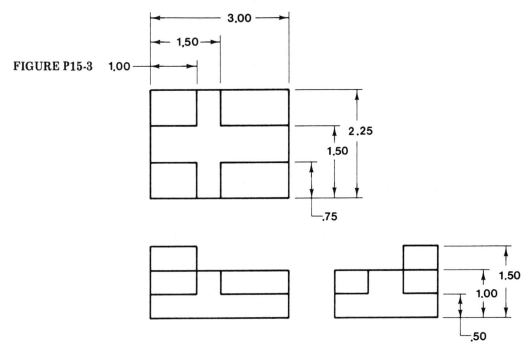

FIGURE P15-3

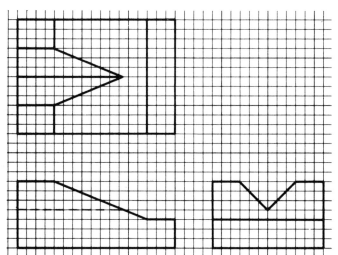

FIGURE P15-4

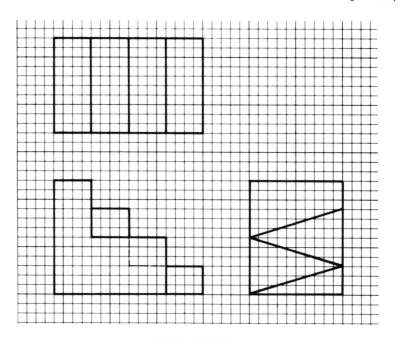

FIGURE P15-5

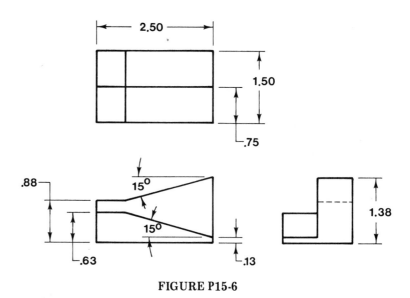

FIGURE P15-6

FIGURE P15-7

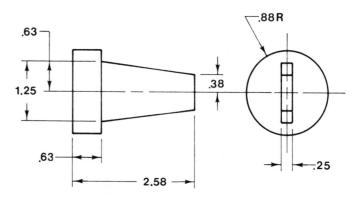

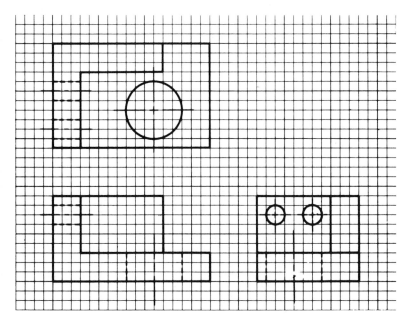

FIGURE P15-8

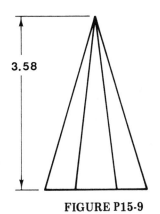

1.13R

FIGURE P15-10

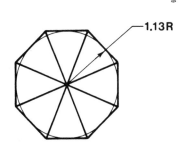

3.58

FIGURE P15-9

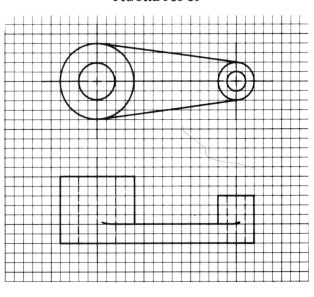

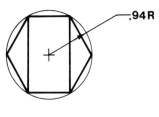

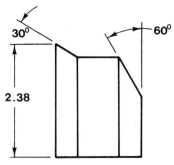

FIGURE P15-11

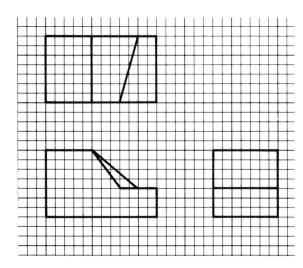

FIGURE P15-12

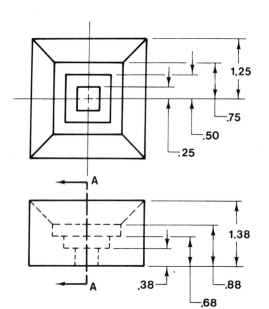

FIGURE P15-13

Object is symmetrical
about both
centerlines

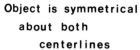

FIGURE P15-14

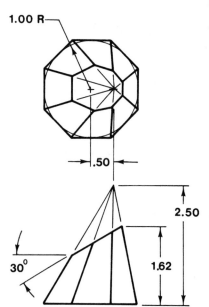

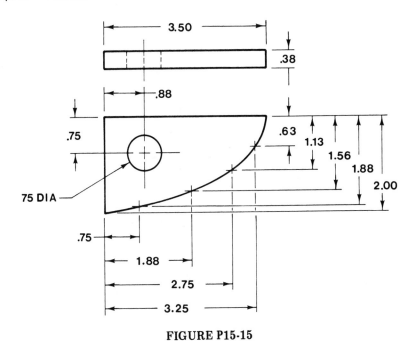

FIGURE P15-15

P15-16 Draw an exploded isometric drawing of the assembly shown in Figure P15-16(a). The washers are 1.00 O.D., 0.50 I.D., and 0.06 thick. Figure P15-16(b) illustrates how to draw an isometric representation of a bolt.

FIGURE P15-16

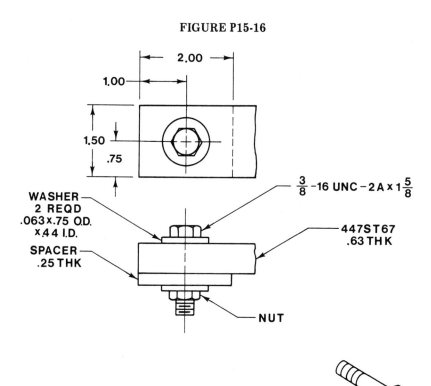

P15-17 Create an isometric drawing of the object shown in Figure P15-17.

FIGURE P15-17

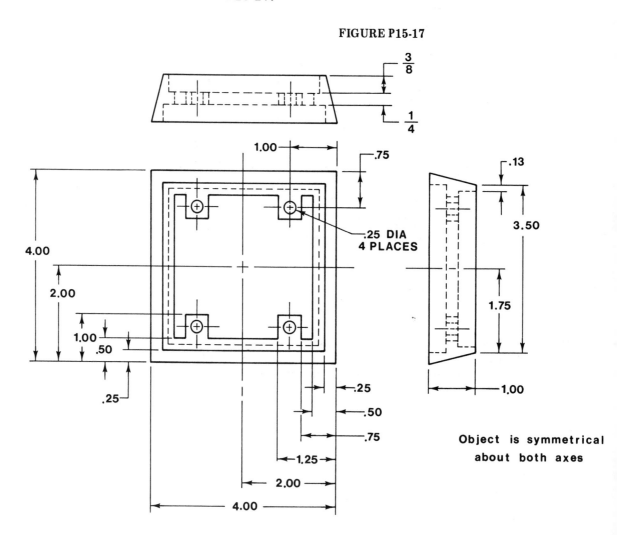

.25 DIA
4 PLACES

Object is symmetrical about both axes

P15-18   Prepare an isometric drawing of the bracket shown in Figure P15-18.

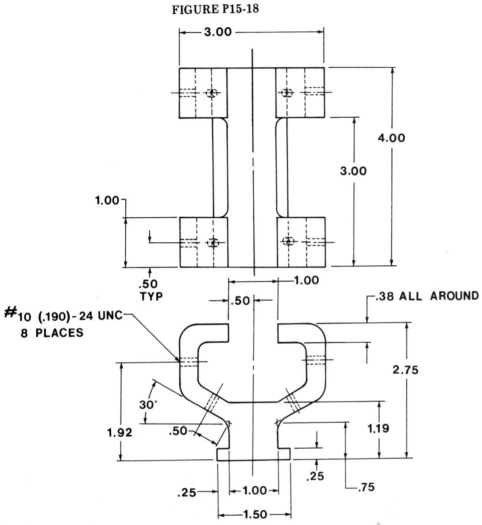

FIGURE P15-18

NOTE: ALL OUTSIDE FILLETS AND ROUNDS $= \frac{3}{8}$ R

ALL INSIDE FILLETS AND ROUNDS $= \frac{3}{16}$ R

P15-19   Metric. Prepare an exploded drawing of the handle assembly
shown in Figure P15-19.

FIGURE P15-19

**HANDLE
ASSEMBLY**

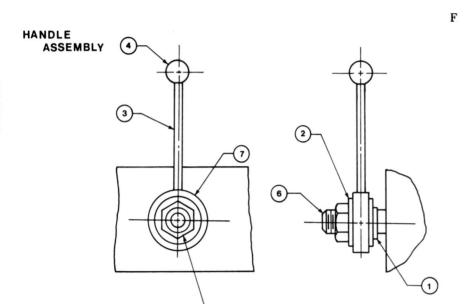

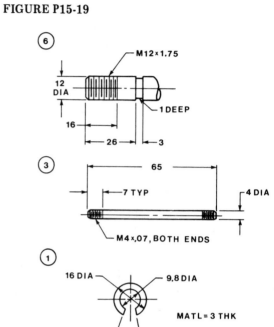

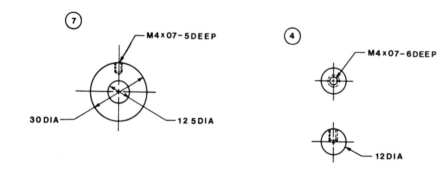

7	HOLDER	1
6	SHAFT, DRIVE	1
5	M12 × 1.75 NUT	1
4	BALL, HANDLE	1
3	SHAFT, HANDLE	1
2	2 × 12 ID × 24 OD	2
1	SNAP RING	1
NO	NAME	QTY

P15-20   Prepare an exploded drawing of the stop assembly shown in
Figure P15-20.

FIGURE P15-20

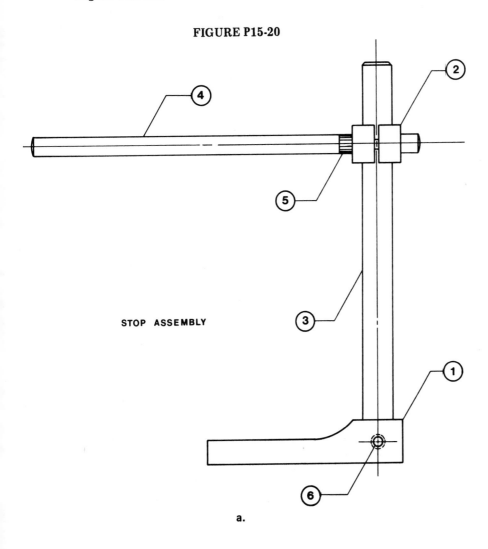

STOP ASSEMBLY

a.

① BASE

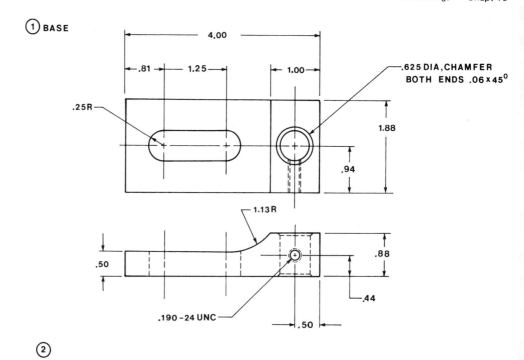

②

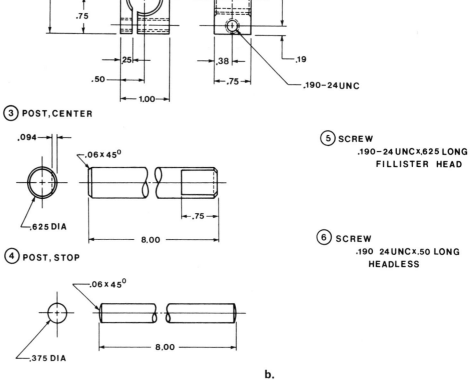

③ POST, CENTER

④ POST, STOP

⑤ SCREW
.190–24 UNC x.625 LONG
FILLISTER HEAD

⑥ SCREW
.190 24 UNC x.50 LONG
HEADLESS

b.

FIGURE P15-20   Continued

P15-21   Prepare an exploded drawing of the scissor jack shown in Figure P15-21.

FIGURE P15-21

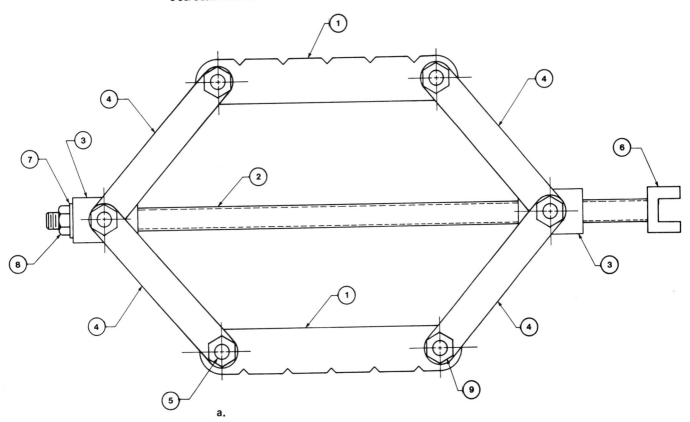

a.

b.

NO	NAME	QTY
9	.31-18 UNC NUT	24
8	.38-16 UNC NUT	1
7	.06 x .38 ID x .75 OD NYLON WASHER	1
6	END PIECE	1
5	HOLDER STUD	12
4	SUPPORT CLIPS	8
3	SLIDER BLOCK	2
2	DRIVE SCREW	1
1	BASE PLATE	2

c.

HOLDER STUD

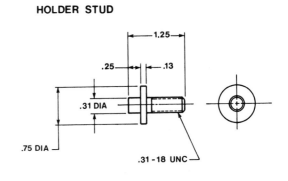

SUPPORT CLIP

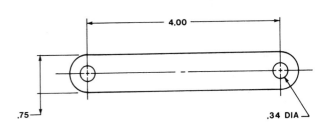

DRIVE SCREW

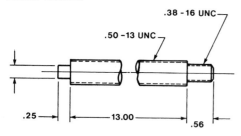

.38 -16 UNC

.50 -13 UNC

.25   13.00   .56

END PIECE

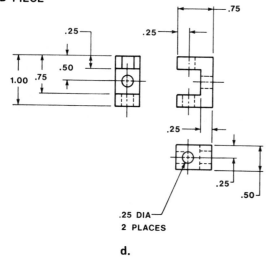

.25   .75

.25   .25

.50

1.00   .75

.25

.25   .50

.25 DIA
2 PLACES

d.

SLIDER BLOCK

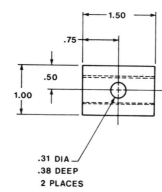

1.50

.75

.50

1.00

.31 DIA
.38 DEEP
2 PLACES

f.

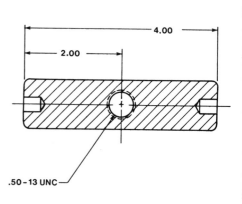

4.00

2.00

.50 - 13 UNC

e.

BASE PLATE

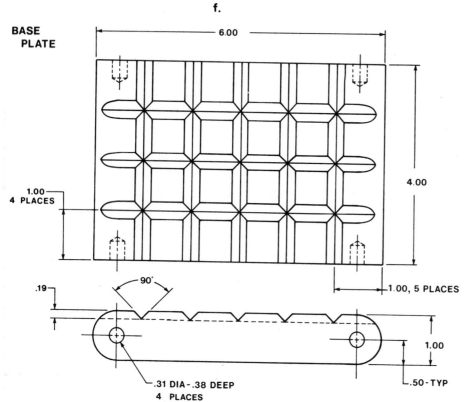

6.00

4.00

1.00
4 PLACES

90°

.19

1.00, 5 PLACES

1.00

.50 - TYP

.31 DIA - .38 DEEP
4 PLACES

FIGURE P15-21   Continued

P15-22   Prepare an exploded drawing of the grinding wheel assembly
shown in Figure P15-22.

FIGURE P15-22

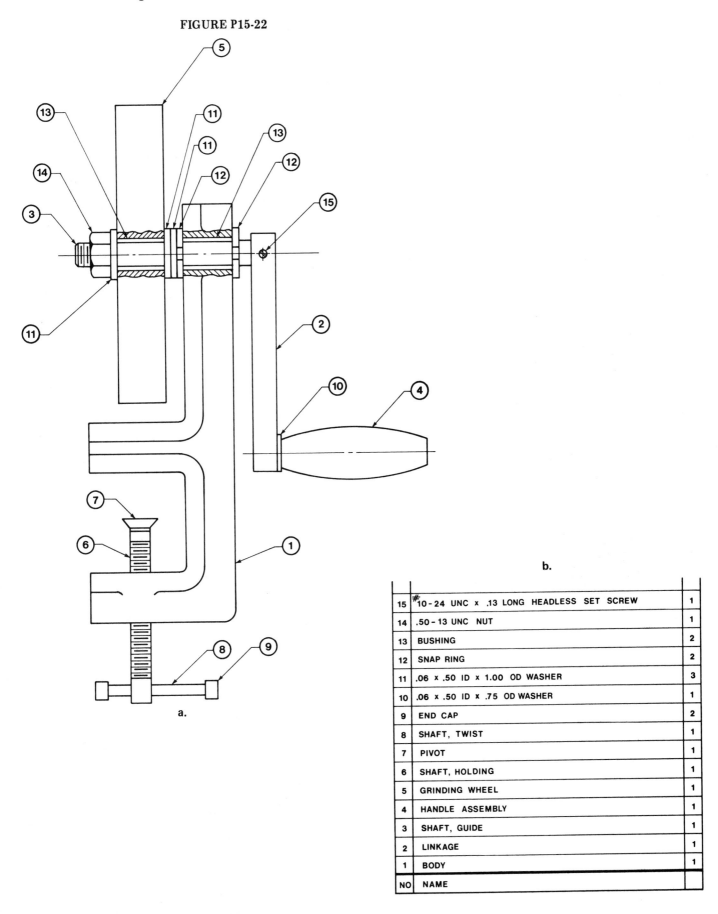

a.

b.

15	#10-24 UNC x .13 LONG HEADLESS SET SCREW	1
14	.50-13 UNC NUT	1
13	BUSHING	2
12	SNAP RING	2
11	.06 x .50 ID x 1.00 OD WASHER	3
10	.06 x .50 ID x .75 OD WASHER	1
9	END CAP	2
8	SHAFT, TWIST	1
7	PIVOT	1
6	SHAFT, HOLDING	1
5	GRINDING WHEEL	1
4	HANDLE ASSEMBLY	1
3	SHAFT, GUIDE	1
2	LINKAGE	1
1	BODY	1
NO	NAME	

BODY

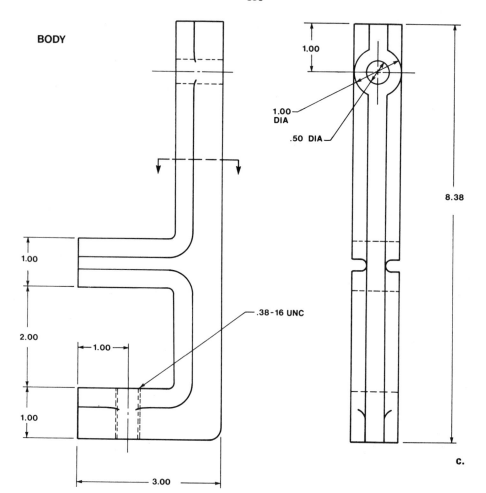

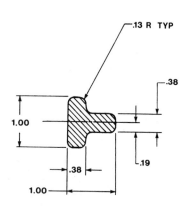

.13 R TYP

c.

d.

HANDLE ASSEMBLY

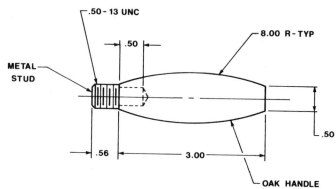

.50-13 UNC

METAL STUD

8.00 R-TYP

OAK HANDLE

SNAP RING
SCALE: 2 = 1

FIGURE P15-22   Continued

.43 DIA

1.00 DIA

60°

LINKAGE

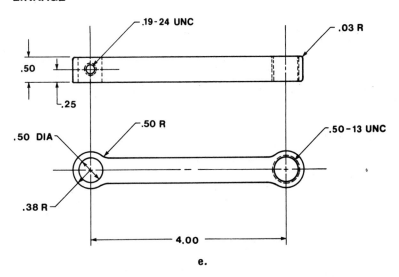

e.

TWIST SHAFT

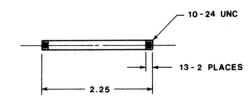

END CAP   Scale: 2 = 1

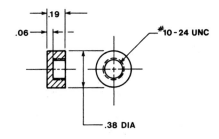

BUSHING

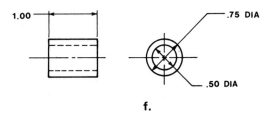

f.

FIGURE P15-22   Continued

HOLDING SHAFT

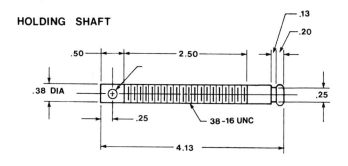

GUIDE SHAFT

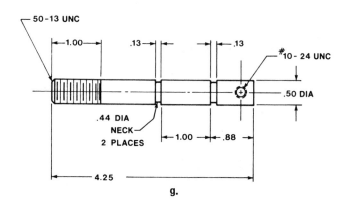

g.

GRINDING WHEEL
SCALE: 1/2 = 1

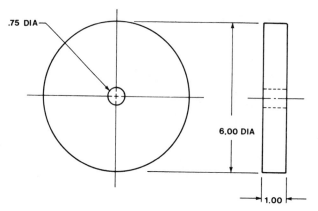

END CAP
SCALE: 2 = 1

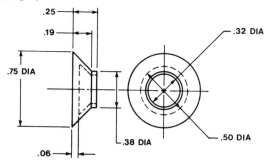

FIGURE P15-22 Continued   h.

P15-23   Prepare an exploded drawing of the pump shown in Figure
P15-23.

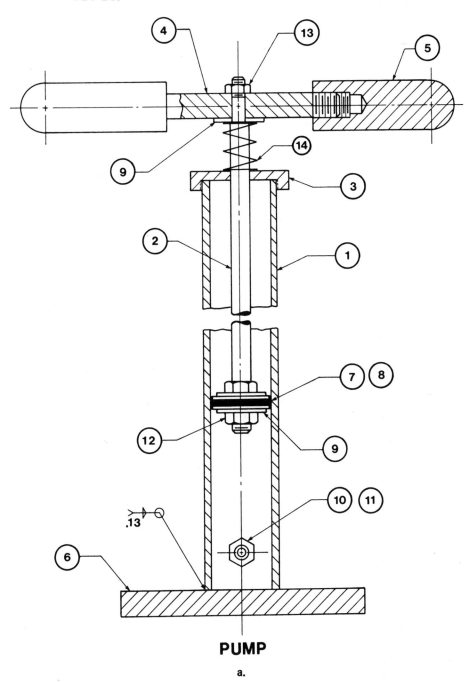

PUMP

a.

FIGURE P15-23

① BODY

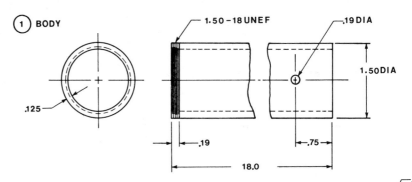

② SHAFT, PUSHER

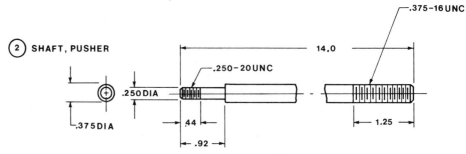

③ CAP

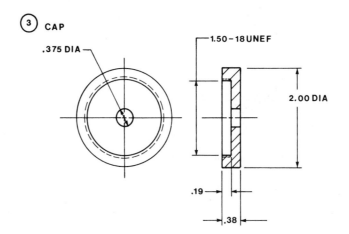

④ SHAFT, HANDLE

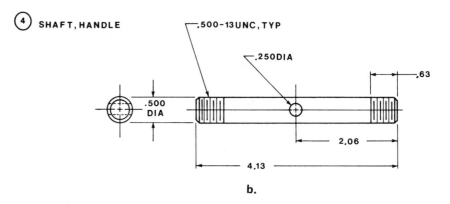

b.

FIGURE P15-23 Continued

⑤ HANDLE

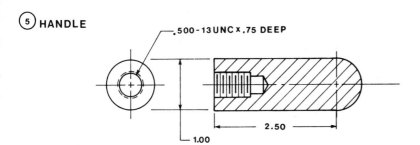

⑥ PLATE, BASE

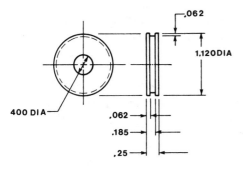

⑦ HOLDER, GASKET

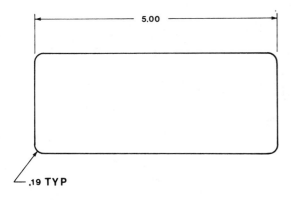

⑧ GASKET

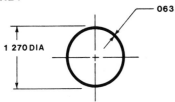

⑨ WASHER

.063 × .38 ID × 1.00 OD

c.

FIGURE P15-23   Continued

d.

⑩ NOZZLE (VENDOR ITEM)

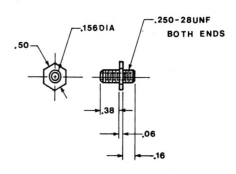

⑪ NUT, CROWN

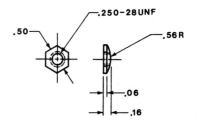

⑫ NUT

.375 – 16 UNC

⑬ NUT

.250 – 20 UNC

⑭ SPRING

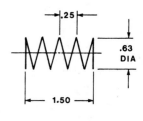

WIRE GAGE = 14 (.064)

FINDING THE TRUE LENGTH OF A LINE BY USING THE REVOLUTION METHOD

1. Define the line in at least two different orthographic views [Figure A-1(a)]. In any one of the views revolve the line so that it becomes parallel to one of the principal plane lines [Figure A-1(b)].

   In the accompanying illustration a line was drawn in the top view parallel to the horizontal principal plane line through point 1_T and then point 2_T was rotated about point 1_T until it intersected this line. The intersection of the line parallel to the principal plane line and the rotation of point 2_T was labeled point $2'_T$.

2. Project the point rotated in step 1 into the other orthographic view so that it intersects a line drawn parallel to the principal plane line through the other view of the point. A line drawn from this point to the nonrotated point is the true length of the line [Figure A-1(c)].

FIGURE A-1

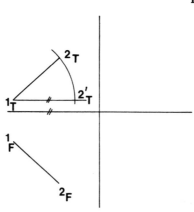

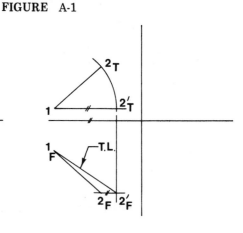

a.   b.   c.

∕ Indicates parallel lines

In the accompanying illustration a line was drawn parallel to the principal plane line through point 2_F and point $2'_T$ was projected into the front view so that it intersected the parallel line as shown. This intersection was labeled $2'_F$. Line 1_F-$2'_F$ is the true length of the line.

Figures A-2 and A-3 are further examples of the revolution method used to find the true length of a line.

FIGURE A-2

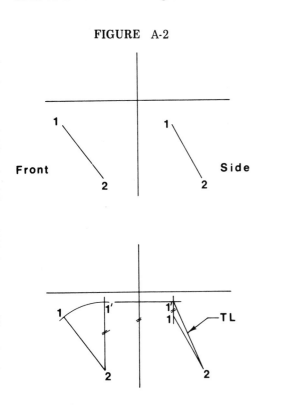

FIGURE A-3

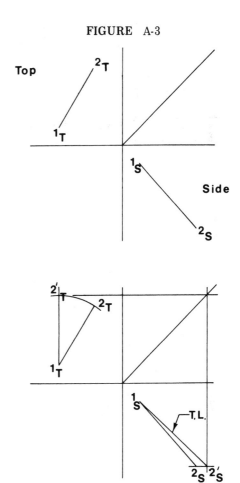

GAME PROBLEMS

This section has been included just for the fun of it. Like most skilled people, drafters enjoy games and puzzles that test and challenge their expertise; thus they often try to stump one another with game problems. Three have been included here (Figure B-1) for you to test your skill. Try them and if you get stuck, write me and I'll send you the answers. Have fun!

FIGURE B-1

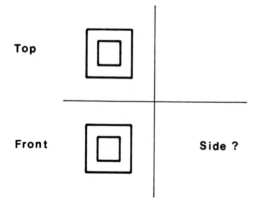

Both are solid objects

Hollow object

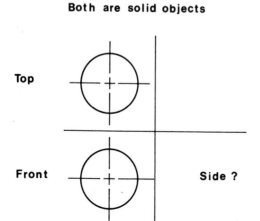

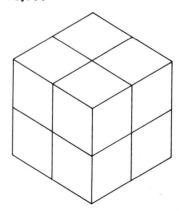

DRAFTING ART

This section presents samples of drafting art. Drafters may use their skill in geometric construction, line technique, and depth visualization to create anything from geometric design to illustration. They are limited only by their own imaginations and aggressiveness. Look over the examples presented in Figures C-1 through C-5 and try copying a few. Then make up your own creations.

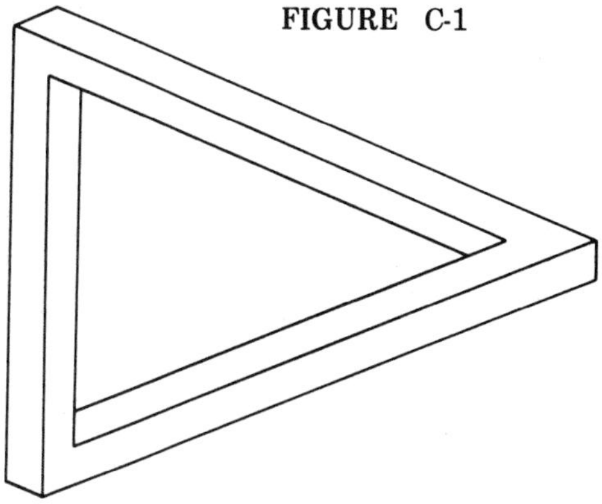

FIGURE C-1

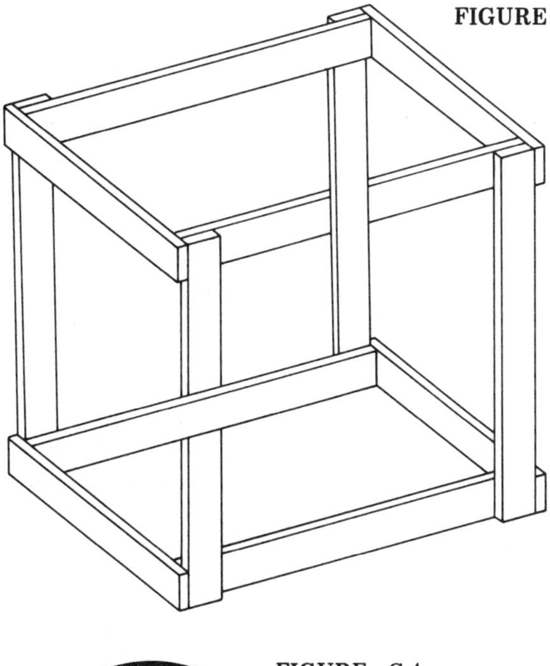

FIGURE C-3

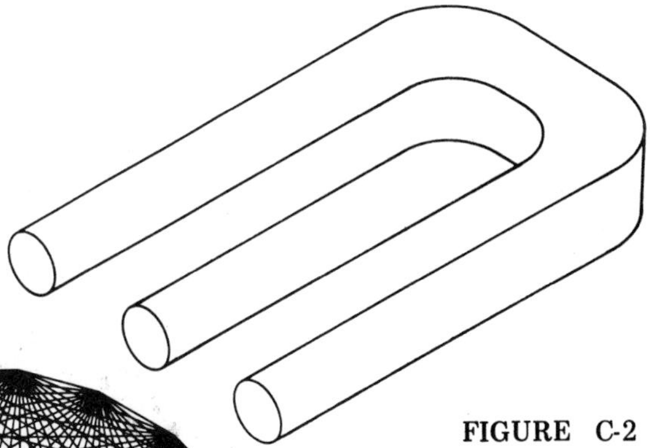

FIGURE C-2

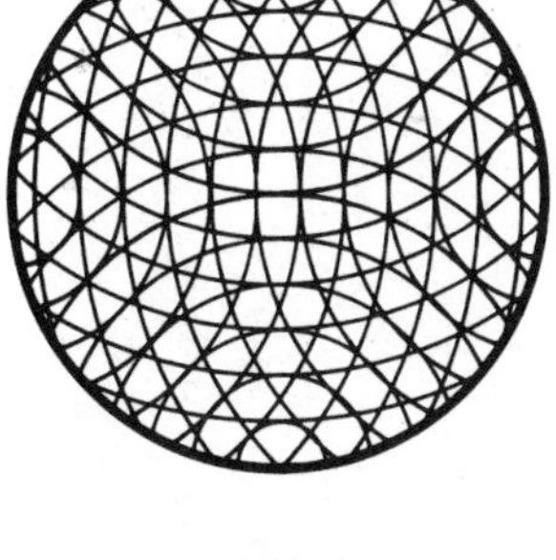

FIGURE C-4

FIGURE C-5

STANDARD THREAD SIZES

Whenever possible, drafters should call for standard thread sizes in their designs. Standard threads may be purchased from many different manufacturers, are completely interchangeable and are relatively inexpensive when compared to "special" thread sizes.

Tables D-1 and D-2 are the UNC and UNF standards. To find the standard size for a given diameter, look up the diameter under the desired thread (UNC or UNF) and read the standard thread size adjacent to it. For example, a 1/4-diameter thread UNC has 20 threads per inch. The drawing callout would be

$$\frac{1}{4}\text{-20 UNC}$$

A 1-1/4 UNF has 12 threads per inch and would be called out on a drawing as

$$1\frac{1}{4}\text{-12 UNF}$$

The size numbers at the top of the tables are for small diameter threads. For example, a No. 4 UNF has a diameter of 0.112 and 48 threads per inch. The drawing callout would be

$$\#4(0.112)\text{-48 UNF}$$

Tables D-3 and D-4 define the 8 and 12 National (N) series thread. In each case, all diameters in the series are made with the same number of threads. All 8 series threads have 8 threads per inch. All 12 series threads have 12 threads per inch. For example, a 1-7/8 diameter series 8 thread would have the drawing callout

$$1\frac{7}{8}\text{-8 UN}$$

TABLE D-1 UNC

Diameter		Threads Per inch (P)
1	(0.073)	64
2	(0.086)	56
3	(0.99)	48
4	(0.112)	40
5	(0.125)	40
6	(0.138)	32
8	(0.164)	32
10	(0.190)	24
12	(0.216)	24
1/4		20
5/16		18
3/8		16
7/16		14
1/2		13
1/2		12
9/16		12
5/8		11
3/4		10
7/8		9
1		8
1 1/8		7
1 1/4		7
1 3/8		6
1 1/2		6
1 3/4		5
2		4 1/2
2 1/4		4 1/2
2 1/2		4
2 3/4		4
3		4
3 1/4		4
3 1/2		4
3 3/4		4
4		4

TABLE D-2 UNF

Diameter		Threads Per inch (P)
0	(0.060)	80
1	(0.073)	72
2	(0.086)	64
3	(0.099)	56
4	(0.112)	48
5	(0.125)	44
6	(0.138)	40
8	(0.164)	36
10	(0.190)	32
12	(0.216)	28
1/4		28
5/16		24
3/8		24
7/16		20
1/2		20
9/16		18
5/8		18
3/4		16
7/8		14
1		12
1 1/8		12
1 1/4		12
1 3/8		12
1 1/2		12

TABLE D-3 Series 8

Diameter	Threads Per inch (P)
1 1/8	8
1 1/4	8
1 3/8	8
1 1/2	8
1 5/8	8
1 3/4	8
1 7/8	8
2	8
2 1/8	8
2 1/4	8
2 1/2	8
2 3/4	8
3	8
3 1/4	8
3 1/2	8
3 3/4	8
4	8
4 1/4	8
4 1/2	8
4 3/4	8
5	8
5 1/4	8
5 1/2	8
5 3/4	8
6	8

TABLE D-4 Series 12

Diameter	Threads Per inch (P)
1/2	12
5/8	12
1 1/16	12
3/4	12
1 3/16	12
7/8	12
15/16	12
1 1/16	12
1 3/16	12
1 5/16	12
1 7/16	12
1 5/8	12
1 3/4	12
1 7/8	12
2	12
2 1/8	12
2 1/4	12
2 3/8	12
2 1/2	12
2 5/8	12
2 3/4	12
2 7/8	12
3	12
3 1/8	12
3 1/4	12
3 3/8	12
3 1/2	12
3 5/8	12
3 3/4	12
3 7/8	12
4	12
4 1/4	12
4 1/2	12
4 3/4	12
5	12
5 1/4	12
5 1/2	12
5 3/4	12
6	12

Table D-5 lists pilot drill sizes for coarse and fine threads.

TABLE D-5 Pilot Drill Sizes for Coarse and Fine Threads
(from ANSI B1.1-1960)

Nominal Thread Diameter		Coarse (UNC, NC)		Fine (UNF, NF)	
		Threads Per Inch	Pilot Drill Diameter	Threads Per Inch	Pilot Drill Diameter
.073	1	64	No. 53	72	No. 53
.086	2	56	No. 50	64	No. 50
.099	3	48	No. 47	56	No. 45
.112	4	40	No. 43	48	No. 42
.125	5	40	No. 38	44	No. 37
.138	6	32	No. 36	40	No. 33
.164	8	32	No. 29	36	No. 29
.190	10	24	No. 25	32	No. 21
.216	12	24	No. 16	28	No. 14
.250	1/4	20	No. 7	28	No. 3
.3125	5/16	18	F	24	I
.375	3/8	16	5/16	24	Q
.4375	7/16	14	U	20	25/64
.500	1/2	13	27/64	20	29/64
.5625	9/16	12	31/64	18	33/64
.625	5/8	11	17/32	18	37/64
.750	3/4	10	21/32	16	11/16
.875	7/8	9	49/64	14	13/16
1.000	1	8	7/8	12	59/64
1.125	1 1/8	7	63/64	12	1 3/64
1.250	1 1/4	7	1 7/64	12	1 11/64
1.375	1 3/8	6	1 13/64	12	1 19/64
1.500	1 1/2	6	1 21/64	12	1 27/64

The thread sizes listed in Table D-6 are *preferred* sizes. Other sizes are available, but the sizes listed in this table should be given preference.

TABLE D-6 Preferred Metric Thread Sizes

Nominal Diameter (mm)	Pitch		Nominal Diameter (mm)	Pitch	
	Coarse	Fine		Coarse	Fine
1.6	0.35	—	16	2	1.5
2	0.4	—	20	2.5	1.75
2.5	0.45	—	24	3	2
3	0.5	—	30	3.5	2
4	0.7	—	36	4	3
5	0.8	—	42	4.5	3
6	1	—	48	5	3
8	1.25	1	—	—	—
10	1.5	1.25	—	—	—
12	1.75	1.25	—	—	—

BIBLIOGRAPHY

Beakley, George C., and Ernest G. Chilton, *Introduction to Engineering Design and Graphics*. New York: Macmillan, 1973.

Brown, Walter C., *Drafting for Industry*. South Holland, Ill.: Goodheart-Willcox, 1974.

Dobrovolny, Jerry S., and O'Bryant, David C., *Graphics for Engineers*, 2nd ed. New York: Wiley, 1984.

Earle, James H., *Design Drafting*. Reading, Mass.: Addison-Wesley, 1972.

French, Thomas E., and Charles J. Vierck, *Engineering Drawing and Graphic Technology*, 11th ed. New York: McGraw-Hill, 1972.

Fryklund, Verne C., and Frank R. Kepler, *General Drafting*. Bloomington, Ill.: McKnight and McKnight, 1969.

Giachino, J. W., and H. J. Beukema, *Engineering Technical Drafting and Graphics*, 3rd ed. Chicago: American Technical Society, 1972.

Giesecke, Frederick E., et al., *Technical Drawing*, 6th ed. New York: Macmillan, 1974.

Grant, Hiram, E., *Engineering Drawing*. New York: McGraw-Hill, 1962.

Hammond, Robert H., et al., *Engineering Graphics*. New York: Ronald Press, 1964.

Hoelscher, Randolph P., et al., *Basic Drawing for Engineering Technology*. New York: Wiley, 1964.

Hornung, William J., *Mechanical Drafting*. Englewood Cliffs, N.J.: Prentice-Hall, 1957.

Jensen, Cecil H., and Helsel, Jay D., *Engineering Drawing and Design*, 3rd ed. New York: McGraw-Hill, 1985.

Jensen, C. H., and F. H. S. Mason, *Drafting Fundamentals*, 2nd ed. New York: McGraw-Hill, 1967.

Luzzadder, Warren J., *Fundamentals of Engineering Drawing*, 6th ed. Englewood Cliffs, N.J.: Prentice-Hall, 1971.

McCabe, Francis T., et al., *Mechanical Drafting Essential*. Englewood Cliffs, N.J.: Prentice-Hall, 1967.

Nelson, Howard C., *A Handbook of Drafting Rules and Principles*. Bloomington, Ill.: McKnight and McKnight, 1958.

Spence, William P., *Engineering Graphics*. Englewood Cliffs, N.J.: Prentice-Hall, 1984.

INDEX